Solving
Riddle of Cancer

Solving the Riddle of Cancer

NEW GENETIC APPROACHES TO TREATMENT

Amil Shah M.D.

Hounslow

Publishers: Kirk Howard & Anthony Hawke
Editor: Shirley Knight Morris
Printer: Metrolitho Inc., Quebec
Front Cover: Histological sections of a tumour model-spheroid-showing fluorescent
stains which identify low pH (green) or proliferating cell location (blue).
Photograph kindly provided by Dr. Ralph Durand.

Canadian Cataloguing in Publication Data:
Shah, Amil, 1949-
       Solving the riddle of cancer: new genetic approaches to treatment

Includes bibliographical references.
ISBN 0-88882-165-4

1. Cancer - Gene therapy. 2. Cancer - Genetic aspects. I. Title.

RC268.4.S53 1994   616.99'406   C94-930451-4

Publication was assisted by the **Canada Council**, the **Ontario Arts Council**, the
Department of Canadian Heritage and the **Ontario Publishing Centre** of the **Ontario
Ministry of Culture, Tourism and Recreation**.

Care has been taken to trace the ownership of copyright material used in this book.
The author and publisher welcome any information enabling them to rectify any references or credit in subsequent editions.

Hounslow Press
A subsidiary of Dundurn Press Limited
2181 Queen Street East, Suite 301
Toronto, Ontario, Canada M4E 1E5

Printed and bound in Canada

CONTENTS

Dedicated to SNS

PREFACE

As recently as the middle of this century, cancer was still a mysterious disease. It seemed to strike with reckless abandon, and once it had gripped its victim, doctors could do little more than relieve the pain, steady the pulse and ease the breathing. It is all too easy now to reflect on this sad state of affairs without realizing that cell biology itself was also a rudimentary science.

We have known about cells for a long time, but the exact molecules that they are made of and the complicated biochemical reactions that bring them to life were largely beyond the capabilities of biologists. Instead, they were obliged to focus their microscopes on the cell and they described in considerable detail its structure. They recognized a multitude of cellular organelles to which they gave fancy names, like Golgi apparatus, without understanding why they were there or what they did.

By the 1950s, things changed. Biologists found out how cells used food to build up new molecules through a series of complex chemical pathways. The discovery of DNA in 1953 was followed by a rapid unravelling of the nature and transmission of genetic information, and the dominant role of DNA in orchestrating all of the cells' activities became firmly established by 1966. But the greatest challenge of biology was yet to come: Trying to understand the essence of life. What was uncovered was an ingenious design, put together over billions of years of evolution. The molecular organization of the cell is immensely subtle and versatile.

The unfolding story of the cell provided an impetus for students of cancer. A concept that guided much of our thinking about the problem arose from the belief that ill-defined changes occurred in the metabolic processes in which the cell is constantly engaged. The popular notion was that the cancer cell conducted its metabolic affairs differently from the normal cell. Finding the differences between the normal and cancer cells was thought to be the key to solving the puzzle. This would allow for selective interference with the cellular functions of the cancer cell while sparing those of the normal cell. The development of a large variety of drugs and the application of high-dose radiation to get rid of the abnormal cells of a cancer were the pragmatic outgrowth of this paradigm. Some success has been achieved in a few types of malignancies but, unhappily, the major human cancers are still on a rampage.

In the past few years, a vastly different view of cancer has emerged. The highly sophisticated tools of genetic engineering have allowed biologists to look deep into the inner provinces of the cell, and what they have learned is taking biology and medicine in a completely new direction.

The root cause of cancer is the malfunction of indigenous genes. The human cell contains a large library of genes that regulate growth. Many of them work in unison to co-ordinate normal development, deciding when a cell should split into two and what type of cell it becomes. But these genes can become damaged, resulting in a deregulation of cell growth that leads to a cancer. In these genes lie new approaches to cancer therapy. The thrust of modern biology is to find ways to redress the genetic damage in cancer cells with the hope to thwart the ravages they inflict.

Only a decade ago, the concept of gene therapy was unknown to most scientists and clinicians. The problems of such therapy were thought to be insurmountable and not given serious consideration. The striking advances in our understanding of cancer in the recent past have, however, changed all of this. It is astonishing how much progress has been made in such a short time; biology has moved from strength to strength, and what seemed daunting not so long ago can now be confidently tackled.

The road to this new understanding of how a cell works and what makes it malignant has not always been easy, but the great achievements are undeniable. With today's cell biology comes the promise of a totally new kind of treatment. I propose to guide the reader in this journey of discovery and share the feeling of marvellous excitement for what is in store for tomorrow.

The disease of cancer will be banished from life by calm,

unhurrying, persistent men and women, working ... in

hospitals and laboratories, and the motive that will

conquer cancer will not be pity nor horror; it will be

curiosity to know how and why.

H.G. Wells. 1927. *Meanwhile.*
George H. Doran Company, New York.

CHAPTER 1

VIRCHOW'S CELL

The origin of the word cancer can be traced back to the Greek *karkinos*, meaning crab. The reason the ancient Greeks chose the crab as the symbol of cancerous diseases is not known with certainty, but whatever it was, the choice seems, in a strange way, a fitting one. The crab, it appears, calls attention to the invasive nature of the disease, a constant and distinguishing feature of cancer. The disease was perceived as possessing a central body, the primary cancer, with claw-like projections, or metastases, extending into and gripping the surrounding healthy tissues of the body. The writing of Paul of Aegina (A.D. 625-690) bears witness to this:

"However, some say that cancer is so called because it adheres with such obstinacy to the part it seizes that, like the crab, it cannot be separated from it without much difficulty."

Cancer is not just a affliction of our time; its history stretches back into antiquity. We find evidence of the disease in fossil bones of extinct animals and in prehistoric mammals. In their summary of the palaeopathology literature, Brothwell and Janssens refer to two tumours in Cretaceous dinosaurs and one in a Pleistocene cave bear, although, admittedly, we must approach this interpretation with due caution as remoteness of time makes it difficult to be quite sure that these abnormal growths were indeed cancerous.

A more reliable record is found in the Ancient Indian Epic, the Ramayana of about 2000 B.C. The Indians described a disease similar in nature to cancer. They carried out surgical removal of tumorous growths, practised cauterization of abnormal tissues, and prescribed an ointment of arsenic compounds.

The dreaded signature of cancer can also be found in the ancient Egyptian papyruses, some of the oldest written records of man. The Edwin Smith Papyrus, a surgical treatise, was written about 1600 B.C., over 3500 years ago. In this, an account of eight cases of "tumours and ulcers" of the breast is given. A fire-drill was used to cauterize the lesions, but failure to control "bulging tumours" by this method was conceded. In the Ebers Papyrus of 1500 B.C., the medical theories of the Egyptians are preserved. This provides further description of superficial "swellings" of the limbs, and warns of the inadequacy of surgical treatment. Most likely, these cancers were diagnosed at a late stage when the disease was beyond the boundaries of surgery, and the admonition that this method of treatment was not to be tried might not have been altogether unwise.

The Egyptians also concocted several recipes for the palliation of cancer. Boiled barley mixed with dates was prescribed for stomach cancer, and a mixture of fresh dates and pig's brain introduced into the vagina was recommended for cancer of the uterus. Perhaps, the most popular remedy was the "Egyptian Ointment", a combination of vinegar and arsenic, which remained in use until the sixteenth century.

Before we judge the ancient Egyptians too harshly, it is sobering to reflect that the remedies for ailments published in a Clinical Guide of 1801 in Edinburgh included digitalis, crabs' eyes, syrup of pale roses, castor oil, opium and "sacred elixir". Even today, it is not difficult to stumble upon "curative" remedies for cancer, promoted in a vituperative manner by practitioners, who are not only scientifically careless but often recklessly dangerous.

Medicine had advanced little until about 500 B.C. in Greece. Here lived Hippocrates, circa 460-377 B.C., the most venerated name in medicine. He rejected many of the widely held superstitions about diseases, and believed that a physical cause could be found for each ailment. His explanation of disease is summarized by William Osler in 1921, thus:

> "The body of man contains in itself blood and phlegm and yellow bile and black bile, which things are in the natural constitution of his body, and the cause of sickness and health. He is healthy when they are in proper proportion between one another as regards mixture and force and quantity, and when they are well mingled together; he becomes sick when one of these is diminished or increased in amounts, or separated in the body from its proper mixture, and not properly mingled with all others."

Hippocrates promulgated the theory of the Four Humours which, he believed, controlled different aspects of the character of the individual. His theory gained wide acceptance and, indeed, seemed consonant with the general cosmogony in which everything was made up of the Four Elements of Air, Fire, Earth, and Water of Empedocles. Normal health required a balance among the four humours. Disease was a generalized condition arising from the imbalance between these. In the Hippocratic view, cancer was a disease of an excess of black bile, or melanchole, produced by the spleen and stomach, but not the liver. So pervasive was his statement of the Four Humours, that the black bile theory of cancer dominated medical thinking for nearly two millennia.

Hippocrates and his school described the behaviour of malignant growths, but it was Aulus Cornelius Celsus, a Roman contemporary of Christ, who is credited with making the clear distinction between cancer, a malignant disease, and other benign growths. Consequently, some of the difficulties in treating cancer became apparent. He realized that limited surgical removal was ineffective for most cancers because of their tendency to recur. This, he suggested, was the unique feature of the disease. Nonetheless, he advised an operation for the less severe types of cancers, such as those of the lip, which remained localized for a long time, and were thus more amenable to complete surgical excision. Celsus wrote an encyclopaedia which included topics in medicine. "De medecina", one of the most comprehensive Latin works on the subject, stands in the transition between Hippocrates and another powerful force in medicine, Claudius Galen.

Galen was born about 129 A.D. in Peragmum in Asia Minor. He studied at Alexandria and travelled widely through the eastern provinces of the Roman Empire before settling in Rome, where he became the most respected physician. He wrote over 500 medical treatises and his views about cancer paralysed original scientific inquiry through the Middle Ages. He based his theories on the doctrine of the Four Humours: blood, phlegm, yellow bile, and black bile. Cancer was thought to develop in those persons in whom black bile was excessive. The black bile was believed to solidify in certain sites, such as the lips, the breast, or the tongue, where it induced the formation of malignant tumours. This was the first conceptual shift away from the idea that cancer was a generalized condition of the whole body, deriving from a lack of balance among the Four Humours.

Despite the inherent fallacy of the black bile theory, Galen appreciated some of the problems of cancer. He drew up a classification in which he described "tumours according to Nature, tumours exceeding Nature, and tumours contrary to Nature." In this last category was placed all malignant tumours.

He understood the virulent behaviour of cancer and the difficulty in controlling the disease. He believed that a cancer arose at a particular site, whence it could spread to other organs. For these reasons, he emphasized that early diagnosis was essential as the cancer could be successfully eradicated if it was still localized. When the disease was more widespread through the body, he advocated systemic therapy, such as purgatives, restricted diets, and blood-letting, and, in the true spirit of a physician, he advised on the use of opium salves for pain caused by advanced cancer.

Between 500 and 1500 A.D., the scientific world was asleep. Galen's doctrine was dogma, espoused by Arabic and European physicians. The tremendous advances in science during this period were not witnessed in medicine. The principal reason for this was that there was scant understanding of the human body. Doctors developed lists of symptoms and could only speculate about the underlying pathology. This led to a lot of guesswork. Not until the Renaissance in Europe did Galen's grip begin to slip.

Andreas Vesalius (1514-1564), a Flemish anatomist, set in motion the forces that would eventually overthrow Galenic anatomy based on the Four Humours. He described what his own dissections of human cadavers revealed, and the first objection to Galenic authority was raised.

In the seventeenth century, William Harvey discovered the circulation of blood. He found that blood does not oscillate back and forth in blood vessels as Galen had proposed, but flows in one direction in a continuous stream from the heart through arteries into veins and back into the heart.

Complementary to the blood circulation is the lymphatic network, first described by Gaspar Asellius of Milan in 1627, and later refined by Jean Pecquet of Montpellier in 1654. As blood traverses the tiny capillaries in tissues, a clear watery fluid, called lymph, exudes through minute openings in their walls and permeates the tissues. Any excess is carried away by another series of small channels, called lymphatics, which eventually connect with the blood stream and empty into it. Nothing remotely resembling the Galenic version of the Four Humours could be verified, and without a firm leg to stand on, the black bile theory toppled.

But as the Galenic black bile theory floundered, new ideas based on abnormalities and stasis of the lymph emerged. The Cartesian sour lymph theory commanded considerable attention. Under normal conditions, extravascular lymph coagulated and evoked a mild reaction; if, however, the lymph fermented and became sour or acidic, then it induced a cancer. The severity of the cancer was attributed to such physical factors as the quality of the lymph, its thickness, and its acidity. Several experiments were performed to confirm or refute the sour lymph theory, but the conduct of these lacked rigorous scientific discipline. In one of the early experiments to marshall support for the theory, cancers were boiled to produce a froth. When lymph, a protein-rich fluid, was coagulated by heating, a froth also resulted. This, of course, proved nothing, but did not dissuade the proponents of the lymph theory to present it as conclusive evidence of the lymphatic origin of cancer.

As the lymph theory ran into difficulties, it was, for a time, rescued by the argument that it was not the lymph itself but its fermentation that was really responsible for the cancer. The explanation was inadequate, but in the absence of any other credible theory, the sour lymph theory flourished and remained at the forefront of medical thinking for nearly 150 years.

The lymph theory was supported by some prominent Parisian surgeons, amongst whose ranks was Antoine Louis, the second director of the Academy of Surgery in Paris. But the most forceful champion was Henri François LeDran. In a memoir published in 1757, he argued so effectively in favour of the lymph theory and against the black bile theory that any lingering credibility of Galen's doctrine was destroyed.

A different approach was offered in 1773 with the sponsorship of a thesis discussion "What is cancer?" by the Academy of Lyon. This attracted a brilliant submission by Bernard Peyrilhe, who was reading for his Ph.D. degree in chemistry at the time. With a pre-vision that is astounding, he called for a new way to tackle the problem. He pointed to four new directions to conduct a systematic investigation of the origin of cancer: the identification of its "toxin", the source of the toxin, the way it acted, and the methods of treatment.

To an extent, Peyrilhe believed in the Cartesian lymph theory, although he recognized that the fermentation of the lymph could not be the full explanation. He considered the possibility of a cancer "toxin", and conducted an experiment to test his hypothesis. He made a cut on the back of a dog, into which he injected extracts from a breast cancer of a woman. He hoped to determine whether the cancer could be transmitted, presumably by its toxin, to the animal. Unfortunately, the experiment was not brought to a successful conclusion, because, so the story goes, his landlady, who could not stand the howling of the dog, drowned it. So ended what might have been a profitable inquiry into the nature of cancer.

Medical progress remained sluggish. Doctors on the whole lacked any drive to delve into the cause of the ailments they saw, but were more satisfied to promote themselves by the manner of the relationship they managed to strike up with their patients. Curiously, the path that led to an improved understanding of diseases, the anatomy of the human body, was open to one group of medical practitioners, the surgeons. But surgeons were not of the same social status as physicians, and the two groups never mingled socially or professionally. The surgeon's attendance to the sick was severely restricted, so whatever knowledge he might have of the anatomical basis of disease, it had little influence in advancing medical knowledge.

In these circumstances, an impasse was reached. The details of human anatomy, so vital for any understanding of diseases, were unavailable. Not until the Italian anatomist, Giovanni Battista Morgagni (1682-1771), undertook a detailed pathological description of the internal organs was methodical study of cancer possible. His chief contribution was the correlation of the clinical history of patients with the post-mortem findings, particulary in cases of cancer of the breast, stomach, bowel, and pancreas. Morgagni set a new high standard of medical inquiry, and his diligent work prepared the way for others who followed.

By the late eighteenth century, an important medical change occurred, brought about through an ironic twist of fate by the political upheaval of the French Revolution. The surgeons, who until then were regarded as simple craftsmen, enjoyed a boost in their esteem and came of age as true medical professionals. One of the first surgically trained chiefs at the new Ecole de Sante was Philippe Pinel, who encouraged a new way of looking at illness. Pinel stressed that a proper understanding of disease could come about only through the careful observation and recording of symptoms, which could then be traced back to the dysfunction of the organs of the body. The old concept of sickness as a single entity with different symptoms depending on the patient's individual pattern of bodily factors was untenable.

The nineteenth century was greeted by significant advances in the understanding of the pathological nature of disease, beginning with the work of the French physiologist and anatomist, Marie-François-Xavier Bichat, a pupil of Pinel. So impressed was he with the organization and the different textures of the organs he encountered in dissecting the human body, he set out to discover all he could about human tissues. In "Treatise on Membranes" published in 1800, he identified twenty-one different types of tissues. (Subsequent detailed microscopic studies refined his classification, and four basic tissues are distinguished: epithelial tissue like the skin, connective tissue like the bone, nervous tissue like the brain, and muscular tissue.) Bichat believed that all diseases were caused by tissue changes. He correctly recognized disease as a localized condition that began in specific tissues.

The importance of Bichat's theory about tissues was largely instrumental in the rise to prominence of hospital doctors. Diseases were now defined in terms of specific lesions in various tissues, and this lent itself to a system of classification and a list of diagnoses. As more clinical observations and experience accumulated, doctors were in a position to prescribe specific therapy based on more accurate diagnosis. The era of empiric therapy, each tailored to suit the patient's individual need and preference, was left behind.

With the new order in France, Bichat's description of diseases was to gain ready confirmation. When a patient died, the corpse would be taken to the dissecting rooms where a post-mortem was performed to find out the cause of death. Relatives of the deceased could object to this practice only if they could come up with the substantial sum of sixty francs for burial.

Bichat himself died at the early age of 31 before he could witness his influence on the direction of medical science. Within 30 years of the publication of his seminal work, another major event occurred that refined his discovery of tissues. In 1829, a London wine-merchant, Joseph Jackson Lister, invented the achromatic microscope. The problem with the earlier microscopes was that the images were fuzzy and unclear because of the asymmetric bending of light rays by the crude lens. Lister was able to design a better lens that did away with these aberrations and provided a clear image. The advent of better microscopes marked the beginning of a steady progress in understanding the tissues that make up the human body.

In 1665, Robert Hooke had examined a thin slice of cork and observed it was made up of tiny compartments. Thinking they were empty, he called them cells. Similar holes were also described in the late seventeenth century in plant tissues by Marcello Malpighi who called them little sacs. The nature or reasons for these curious holes in tissues were unknown. In 1809, however, Gottfried Treviranus separated out the cells of a buttercup, and demonstrated that they were discrete entities, although their function remained unclear.

The first step towards understanding the cell came in 1831 with the observation, under the newly developed achromatic microscope, of the cell nucleus by the German botanist, Mathias Schleiden. In a casual after-dinner conversation, Schleiden described his studies of the microscopic structure of plants to the zoologist, Theodor Schwann, who immediately seized upon the similarity between plant structures and his own observations in animal tissues. A comparison of cells from specimens of plants and animals convinced both men of the remarkable similarity of cells in members from both kingdoms. Schwann decided to examine every kind of tissue known to him. He summarized his observations in 1839 in a paper entitled, "Microscopic studies in the correspondence in the structure and growth of animals and plants". With forceful clarity, he wrote:

"The great barrier between the animal and vegetable kingdoms, namely, the diversity of ultrastructure, thus disappears."

The recognition that all plant and animal cells were essentially similar was a milestone in biology: The cell theory was enunciated.

Schwann found that cells were grouped differently in different tissues. For instance, in blood or lymph, the cells were free, whereas in bone they were held together by an intercellular substance. At the same time, Johannes Müller, who was the mentor of Schleiden and Schwann, was engaged in the study of a reclassification of large collections of tumours in the museums of Berlin, Halle, Brunswick, and London. He distinguished the various types of cancers microscopically, and noted that they were frequently composed of "primitive" or immature cells arranged in a disorganized manner. In 1838, he published his important work, "Ueber den feinern Bau und die Formen der Krankhaften Geschwülste — On the Nature and Structural Characteristics of Cancer". His awareness that all cancers are made up of cells makes his one of the monumental contributions to biology and cancer research.

Things moved at a rapid pace. Schwann had demonstrated that cells were the structural element of Bichat's tissues, and in them must, therefore, reside the very essence of life. In the same year, Jan Evangelista Purkinje described a jelly-like substance in animal ova and embryonic cells. He called it protoplasm, reflecting his belief that in this half-solid, half-liquid material could be found the ingredients of life.

The discovery of cells and their universal presence in vegetable and animal tissues raised the intriguing question of their origin. Schwann had already suggested that cells came into existence either inside or near another cell. But the notion of spontaneous generation, whereby cells were envisaged to arise from formless body fluids, such as lymph, held sway. In 1846, Karl von Baer made an astute observation. He noted that a living sea-urchin egg cleaved into two separate cells, and just before this the nucleus also split into two. This led to the dethronement of the hypothesis of spontaneous generation of cells in 1852 by Robert Remak in his succinct but eloquent aphorism: Omnis cellula e cellula — all cells come from other cells.

The cell theory was advocated more forcibly by Rudolph Ludwig Karl Virchow (1821-1902), of Berlin, one of the great medical figures of all time. In his early life, he was a radical involved in the German revolution of 1848, and his life-long love of truth was his crowning glory in science and in politics. Virchow wrote extensively on several topics in pathology, parasitology, anthropology, as well as public health and politics.

His diligent and careful work on the cell was his chief medical contribution. He wrote, "We can go no farther than the cell. It is the final and constantly present link in the great chain of mutually subordinated structures comprising the human body."

Virchow recognized that cells were different and some specialized in certain functions. He introduced a new way of looking at the body as a cell-state, in which every cell was a citizen. "An organism is a society of living cells, a tiny well-balanced state." In his view, diseases were seen as a civil war between cells.

He was the first to extend the cell theory to diseased tissues, and his notion of cancer was consonant with this general notion of disease. He believed that a cancer arose as a result of a change in character to cells of a different type. Virchow realized that a full understanding of illness and health could not be possible without a consideration of the cell.

Two hundred years ago, the view of diseases was quite different. Each person's ailment was seen as a unique condition, the remedy decided through speculative pathology and guesswork. Medical practice was hopelessly confused. The inquiry started by Bichat at the turn of the nineteenth century and brought to its triumphant maturity by Virchow altered the medical profession in profound ways. An accurate diagnosis for each disease was required and therapy was specific.

All this new understanding of disease and the careful scientific attitude it fostered allowed for the correct diagnosis of cancer and a more rational approach to its cause. With the recognition that cancer was ultimately a disease of cells, a few possibilities regarding its origin seemed immediately obvious. One of the early theories was proposed first by Müller, and later, in a more forceful manner, by Julius Cohnheim. Müller had described that cancer cells were immature, not unlike the cells of an early embryo. In a textbook of pathology published in 1877, Cohnheim put forward the theory of "embryonic rests". He believed that at an early stage in the developing embryo, certain cells were misplaced, forming embryonic rests. These remained quiescent, but if the conditions were right, they could grow into a malignant tumour.

Another interesting theory, propounded by Blair Bell, took the reverse tactic. Bell suggested that a cancer arose as a result of a reversion of the mature, adult cell to the early embryonic cell which had considerably more capacity for growth. He argued that if it was possible for a cell to pass through all the stages of maturation from the embryo to adult, then under certain conditions, it would be possible for the mature cell to retrace the steps back to an embryonic stage. Indeed, as pathologists had long observed that a common feature of cancer cells was their "primitive" appearance, the cell reversion theory seemed plausible. Bell believed that several stimuli — mechanical, bacteriological, radiological, thermal, and others — could induce the cell to undergo retrogressive changes by disturbing its metabolism. The injured cell, in an attempt to survive, reverted to the ancestral type present in the embryo. In this way, its chances of survival were enhanced, as the embryonic cell was perceived as having a wider range of potential, such as invading blood vessels in its search for nutrients.

The modern view of the formation of cancer was ushered in by the German zoologist, Theodor Boveri (1862-1915), one of the founders of the science of genetics, that arose in the first decade of this century. Boveri was an intellectual giant in the laboratory, but was equally at home and competent as a pianist and painter. Boveri took great interest in chromosomes and realized that they were the agents of hereditary information. The full complement of chromosomes was necessary for an organism to develop properly. From his observations of sea urchin eggs which were experimentally induced to divide abnormally, he proposed that a cancer might result from an unbalanced set of chromosomes. The essence of his theory was that malignancy involved a certain abnormal "chromatin complex" or chromosome set. A normal chromosome set was necessary for the normal functioning of a cell; if a single chromosome was absent, the cell would be defective and it would die. He speculated that there might be chromosomes which inhibit cell division; loss of these could lead to unlimited growth. Alternatively, there might be chromosomes which promote cell division; an excess of these would result in rapid proliferation. To Boveri, a delicate chromosomal balance was essential for normal cell function and growth. Any disturbance of this could lead to the birth of a cancer.

What Boveri had done was to portray the cancer cell as a caricature of the normal. In so doing, he dislodged the most important impediment to the scientific understanding of the problem. The seemingly unnatural behaviour of cancer, embodied in the Galenic verdict of these tumours being "contrary to Nature", had imposed a conceptual barrier. Certainly, nothing so devastating and uncontrollable could emerge from a healthy body.

"We can go no farther than the cell," Virchow had said in his view of illness and health. Boveri's work prompted scientists to look at the cell to find out more about cancer, and in this new atmosphere, medicine was free to move towards defining those factors that might bring about this distortion.

CHAPTER 2

THE LANGUAGE OF LIFE

Cells are a triumph of Nature. The living cell is complex and beautiful, an expression of four billion years of evolution. Because life started on earth such a long time ago, it is difficult to imagine what the first living things were like, but certainly the early stirrings were humble. In the oceans and ponds of the early earth's surface, some molecules arose that were able to make crude copies of themselves, using as building blocks other molecules that floated around nearby. No doubt, at first this reproduction was inefficient, but as time went on, variants came into being that were better at making copies. Molecules with specialized functions eventually came together for the better good of all: the first cell was born.

In the process of evolution, creatures with many cells developed from earlier ones having only single cells. But because they are descended from some of the first organisms, all of today's living things — bacteria, plants, animals, and humans — share a common ancestry. We have a lot in common with other living things. The immense variety of nature uses a single language, a testimony to the biochemical unity of life.

The living cell is an elaborate, well-organized chemical factory which takes one set of organic molecules — food — and, in several discrete steps, breaks these down and reassembles them into cellular machinery. The running of the cell's machinery is supervised by a special class of molecules, the nucleic acids, which live in the deep interior of the cell, the nucleus. The typical cell is depicted in Figure 1.

Extending through the nucleus are a multitude of coils and strands, the chromatin network, made up chiefly of DNA. The DNA contains the full catalogue of instructions, collected from all kinds of sources in nature, filed for all kinds of contingencies. More than storing the information on how to make a cell work, the DNA molecules know how to make identical copies of themselves. They are the hereditary material, passed on from one generation to the next. These are molecules of awesome power.

Because DNA is the hereditary material, its accurate and faithful transmission from the parental cell to the daughter cells of the next generation is of critical importance. This is achieved by a special process of cell division called mitosis (Figure 2). The nuclear chromatin network normally forms delicate intertwining fibres and just before the cell divides, it goes through a complex dance of its own. First, each fibre coils into a short rod, the chromosome. In this way, the DNA occupies a very much smaller space than would be necessary were it uncoiled. Each animal or plant has its own number of chromosomes, and always the same. The garden peas have fourteen, fruit flies have eight, some kinds of butterflies have more than three hundred, and humans have forty-six.

Figure I
A typical cell as seen under the microscope

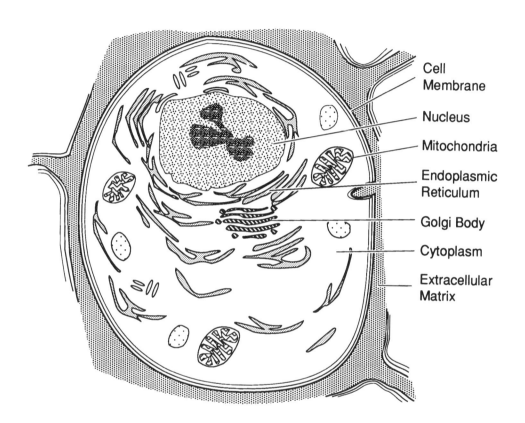

Cell
Membrane

Nucleus

Mitochondria

Endoplasmic
Reticulum

Golgi Body

Cytoplasm

Extracellular
Matrix

Figure 2
Stages of cell division or mitosis

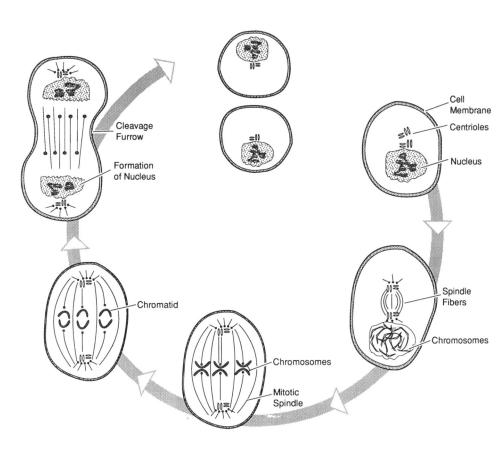

In the initial phase of mitosis, an identical copy of the DNA in each chromosome is made, following which the chromosomes divide into identical halves called chromatids. The chromatids move to the equator of the cell. In the succeeding events, the two identical chromatid sets separate, one set going to one end of the cell, and the other set going to the other end. At this point, a wall begins to build across the middle of the cell, ultimately dividing it into two. What, an hour before, had been one cell is now two, each containing the full complement of genetic DNA.

It is probably evident that the germ cell — the egg of the female or the sperm of the male — must undergo a different kind of cell division. A moment's thought will suggest the reason. If each germ cell contributes the full complement of forty-six chromosomes, the new cell produced by fusion of the egg and sperm would contain ninety-two chromosomes. In fact, germ cells contain only twenty-three chromosomes. This is brought about through a special type of cell division, called meiosis, with two nuclear divisions accompanied by only one division of chromosomes. When fertilization occurs, the full complement of forty-six chromosomes is restored.

An important inference from this is that the human cell contains not forty-six different chromosomes, but twenty-three pairs of chromosomes, one set from the mother and one from the father. Since genes, discrete units of DNA, are located on chromosomes, it follows that each gene is also present in duplicate. One further point is worth mentioning. Each pair of chromosomes lines up at the equator of the cell during meiosis. The manner in which each pair separates and moves to the pole of the cell occurs independently of the other pairs of chromosomes. This results in a random assortment of gene combinations in the germ cells, and accounts for the various "mixes" of genetic characteristics of individuals.

Some of the fundamental principles of heredity were discovered through the clever experiments of Gregor Mendel. Mendel was born in 1822 to a family of Silesian peasants, who lived in a part of the Austrian Empire that is now the Czech Republic. The only escape from the harsh peasant life was through education, and he entered the Augustinian monastery at Brno. The monastery was famous throughout Austria for its botanical museum and gardens, and well-stocked library. He was ordained a priest in 1847, and a few years later was sent to the University of Vienna, where he studied physics, mathematics, and the natural sciences. He presented himself for the examinations that would certify him as a teacher, but ran into a brick wall. His close study of contemporary debates on plant evolution did not go over well with his examiners. Mendel became impatient and abruptly withdrew during an oral examination. But his penchant for independent thinking and a respect for disciplined observations were to reward him later. He returned to Brno in 1854, and soon began an immense and careful experiment of pea plant breeding that would take over fifteen years to complete.

Mendel was interested in the evolution of plants and animals and how the parental generations passed on their traits to the succeeding generations of offspring. It was generally believed at the time that when the different varieties of the same species were cross-bred, the parental traits were diluted or mixed in an unpredictable manner. Mendel did not accept this explanation. He believed that there were laws in Nature, as in all branches of science, which dictated the rules of heredity. He set about to prove this in his famous experiment in the monastery gardens at Brno.

There are few instances in science where one can trace the source of a major discovery as precisely as Mendel's principles of heredity. His clear thinking on the subject is his outstanding contribution to genetics. Mendel's laws, established with the use of the simple pea plant, were later shown to apply to animals and man.

The choice of the garden pea, *Pisum sativum*, was not an accident, but the result of careful observation and thought. Mendel realized that pollination could be easily controlled; the pea plant is self-fertilizing, but cross-fertilization could be readily achieved by merely removing the stamens, containing the pollen organs, from one plant and transferring them to another de-stamenized plant. Also, the pea plant is easy to cultivate, and two crops could be grown in a single season, expediting the conduct of his experiments. Further, he noted that the pea plant has many distinct traits that could be followed through many generations. Mendel chose seven "unit characters" or traits, such as the colour of flowers, shape of seeds, or size of stems, each of which had two alternative appearances. For example, the colour of the flower could be either red or white.

Mendel's experiments on the inheritance of the flower colours serve to illustrate his methods. This is depicted in Figure 3. He selected two varieties that consistently produced flowers that were either red or white through many generations. After crossing these two lines, he kept accurate records of the results. He noted that the first generation of plants grown after the first breeding experiments produced only red flowers. When the plants of this generation were allowed to fertilize themselves, the second generation grew into 705 red and 224 white plants.

Figure 3
Mendelian pattern of inheritance of flower colours in the pea plant

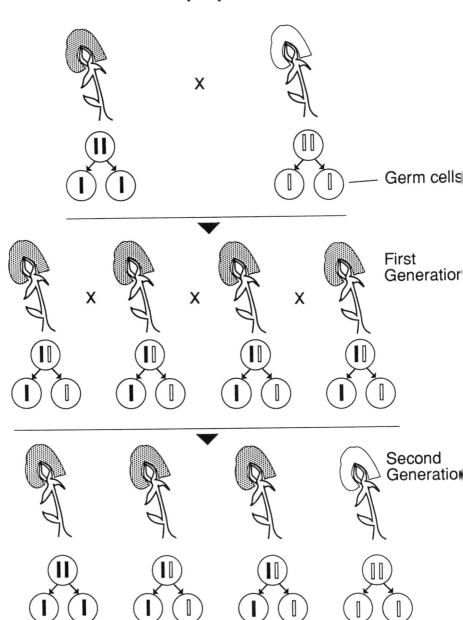

In all cases, crosses between two different varieties for each trait produced offspring that were of one type in the first generation. However, when these plants were permitted to reproduce by cross-fertilization, the traits of the original parents re-emerged.

Mendel had finally unlocked the secret of heredity. First, he suggested that hereditary traits are controlled by units, now called genes, that are passed on to each successive generation. Second, to explain the disappearance and reappearance of a particular trait in a predictable ratio, he correctly deduced that each trait is produced by two genes, with one being contributed by each parent and recombined at fertilization of the male and female germ cells. Third, when two contrasting genes are present, only one is expressed. In Mendel's experiments, the gene for red flowers is said to be dominant, while that for white flowers is recessive. The dominant gene is expressed in the presence of a recessive gene; the recessive gene can be expressed only if present in duplicate.

While Mendel's principles of inheritance remain valid to this day, certain modifications have been introduced with the recognition of some other hereditary patterns. One such pattern is the interaction of genes. We said that genes are either dominant or recessive. Genes, however, interact in such a manner that the end result is an intermediate state. The second pattern of inheritance that does not follow the classic Mendelian mode is the dihybrid cross. In the example of the colour of flowers of the garden pea, one pair of genes was considered; that is, a monohybrid cross. If two hereditary traits are involved, a dihybrid cross, the principles are the same, but two pairs of genes, perhaps even on different chromosomes, must be followed. This exercise can be extended to three or more pairs of genes. Obviously, the calculation of the probability of the different traits showing up in the offspring becomes increasingly complicated, but it does not detract from the general Mendelian laws.

Mendel presented his results to the Brno Society of Natural Science in 1865, and published them in the proceedings of that society a year later. There followed a long silence. One of the reasons for this neglect was that his ideas ran contrary to the prevailing concept that there was a gradual change from one generation to the next, a concept embraced by biologists concerned with the questions of heredity and evolution stemming from Darwin's publication. Rather than support the notion of continuous variation, Mendel's demonstration that variation in plants could be traced back to the original variability of the parents seemed contradictory. Although on a superficial level Mendel's principles appeared at odds with Darwin's evolutionary theory, the possibility that variation in character may occur through small, discontinuous steps, each obeying Mendelian laws, is now generally accepted.

Mendel became absorbed in his duties as the abbot of his monastery and published nothing further on the subject. But by the end of the century, biologists became more interested in the pattern of inheritance of single traits, and this inevitably led to a rediscovery of his classic experiments.

By the late 1800s, the nucleus was demonstrated to be a universal feature of all plant and animal cells. A more detailed study of mitosis showed that chromosomes split lengthwise, with each chromatid migrating to opposite poles of the cell, where it eventually becomes part of the nucleus of the daughter cell. Here then is a mechanism that permits the genetic information in the cell to be passed on fully to the next generation of cells. What is more, the chromosomes are also the vehicles for conveying inheritance. We get half of all our genes from our mother in the egg, and the other half from our father in the sperm.

What was not immediately apparent was how could genes, sequestered in the cell's nucleus, know what to do in the cell, and how do they send these instructions to the rest of the cell. An understanding of the gene's actions came from working out the biochemical reactions that go on within the cell. In 1897, Eduard Buchner discovered enzymes, a class of protein molecules that serve as assembly-line workers, each specializing in a particular task in the cellular factory. At the time, it was thought that the cell must be intact to carry out any biochemical reaction. Buchner was interested in whether cellular fragments could do the same. He crushed yeast cells, and to preserve the juice, he used an old kitchen trick; he added sugar. What resulted was the fermentation of the sugar solution by the yeast juice. Thus, enzymes were discovered. It was not long after that enzymes were found in other types of cells, and each cell contained many distinct kinds of enzymes. Ten years after his discovery of enzymes, Buchner received the Nobel Prize.

Enzymes are protein catalysts in biological systems, and most biochemical reactions in the cell would occur very slowly were it not for catalysis by enzymes. Even a small amount of enzyme can have a large effect. Each enzyme is very specific and catalyzes only one particular reaction. As might be expected, specialized cells contain unique enzymes to aid in performing their special tasks. For example, the red blood cells which transport oxygen contain an abundance of a special protein, haemoglobin, and their cellular machinery is geared towards its production.

The complexity of the cellular manufacturing processes was appreciated in 1941 by George Beadle, a geneticist, and Edward Tatum, a biochemist, at Stanford University, from their experiments with the common bread mould, *Neurospora crassa*. They found that production of a substance in a cell occurs in a series of steps, each of which is catalyzed by an enzyme. For instance, we can imagine a product (P) manufactured from a substance (S) through three intermediate steps, as follows:

$$S \longrightarrow A \longrightarrow B \longrightarrow P$$
$$\text{step a} \quad \text{step b} \quad \text{step c}$$

If any of these steps is interrupted, there is a deficiency of the final product.

Beadle and Tatum postulated that a block in the metabolic pathway can result from a lack of the enzyme that catalyzes any of the steps. Further, since each enzyme is the product of a gene, they coined the famous slogan "one gene — one enzyme".

When they introduced this idea, it was not generally accepted, but later recognition of their work led to their award of a Nobel Prize in 1958.

Through the efforts of Beadle and Tatum, a general scheme of how living things work began to emerge. Each gene controls the construction of a particular protein. Some of these proteins form the scaffold for the cell, while others decide what cellular reactions should take place. Each cell in our body has a complete set of genes in its nucleus, and the way these are switched on or off determines how the cell grows, functions, and, eventually, dies.

This new understanding did not solve the problem of how a

gene worked, but it provided a framework within which the question could be tackled. By the mid-1940s, evidence was mounting that pointed to DNA as the central molecule in the mystery. The first glimpse into this actually came in 1928 through the astute observations of a British bacteriologist, Fred Griffith. He knew that there were two strains of the bacteria that caused pneumonia: the smooth (S) and the rough (R), depending on the presence or absence of a capsule around the bacterial cells. Normally, the progeny of the S strain is also S type, and likewise, the R strain gives rise to R type only.

Griffith was able to make the R strain change into the S. This he did by injecting a mixture of the living R type and dead S type into mice, from which he subsequently recovered living S bacteria. As the dead S bacteria could not grow, and the R bacteria bred true to type, he speculated that a change must have occurred through the transfer of a substance from the S to the R strain that endowed them with the ability to produce a capsule.

Griffith's work was pursued further over the next 15 years by others, who clearly demonstrated that the ability to produce a capsule could be transmitted from the S strain to the R strain of the pneumococcal bacteria. Further, once this ability to produce a capsule was acquired by the R bacteria, it was passed down as a stable heritable trait to its descendants; in other words, it behaved like a gene.

The search for the substance responsible for the bacterial transformation culminated in a comprehensive study by Oswald Avery, Colin MacLeod, and Maclyn McCarty at the Rockefeller Institute in New York. In 1944, they published a paper claiming that a chemical extracted from an encapsulated pneumococcal S strain converted a non-encapsulated R strain, so that the recipient bacteria acquired the capsule. The biologically active ingredient was DNA.

The experiment of Avery and his colleagues was greeted with mixed reactions, but its influence was unquestionable. In 1945, Avery was awarded the Copley Medal by the Royal Society in London for his work on the transforming factor.

Now that Avery had provided compelling evidence that the gene is made up of DNA, the next task was to unravel its chemical structure. The general plan of the DNA molecule is extremely simple. When we reflect that we carry stores of DNA in our nuclei that have come originally from an ancestral cell created in the molecular Garden of Eden that was the earth about four billion years ago, this should not be astonishing.

The DNA molecule has a backbone comprising sugar molecules that are bridged by phosphates to make repeating units (phosphate - sugar - phosphate - sugar) many thousands or even millions of times over. The sugar is deoxyribose; hence, the name deoxyribose nucleic acid or DNA. "Nucleic" refers to its presence in the cell's nucleus, and "acid" refers to the phosphate groups which carry a negative charge.

Each sugar has a side group attached to it. There are four side groups, called bases, which are conveniently denoted by their initial letters — A, G, T, and C for Adenine, Guanine, Thymine, and Cytosine. Because of their size and shape, and their chemical make-up, A will pair neatly with T, and G with C.

The discovery of the three-dimensional structure of DNA depended largely on the technique of x-ray crystallography. Shining x-rays through a crystal produces a pattern of reflections from which the arrangements of atoms within the crystal can be deduced. In 1950, Maurice Wilkins and Roslind Franklin at King's College, London, took the crucial x-ray photographs of DNA molecules that had been carefully prepared in Bern by the Swiss chemist, Rudolf Signer. Dry DNA has the appearance of irregular white fluffs of cotton, but when it takes on water, it can be drawn out into thin fibres. Within these fibres the individual long, thin DNA molecules line up parallel to one another. When such fibres are placed in the path of an x-ray beam, the rays are bent, and from the pattern of the reflections, the arrangements of the atoms can be determined. From this, a precise structure emerged, with the individual bases stacked perpendicular to the long axis of the molecule as if they were a pile of pennies (Figure 4).

From the data obtained by Wilkins, Francis Crick and James Watson built models of the DNA. Others were also doing the same, but Crick and Watson took a novel approach to the problem that was to pay off. They proposed that DNA is a two-chained structure in which the two chains lie together, side by side, twisted around one another to form a double helix, and linked together by their bases. The secret of their success in building the DNA model was the recognition of the significance of the base pairs: A with T, and G with C. The bases are on the inside, and the chains are coiled to permit the bases on one chain to face its partner on the other. A and G are large, while T and C are smaller, so that each pair consists of a large base and a small base, thus providing a dyadic symmetry. The dimensions of the double helix are such that there is insufficient space to permit the two large bases to pair up, and would allow too much space between the two smaller bases.

The chains of the double helix are not identical, but are complementary to each other. This is the result of strict pairing of the bases: A to T, and G to C. Each of the two pairs can be inserted in the double helix in one of two ways: A=T or T=A,

Figure 4
The basic structure of DNA

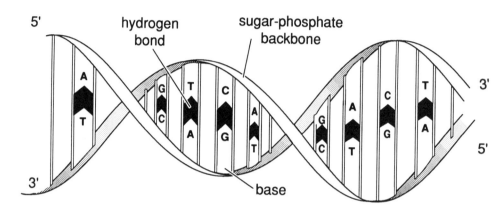

and G≡C or C≡G. The base pairs can exist in any sequence along the DNA chain, but since the pairing is specific, if the sequence of one chain is known, the sequence of the complementary chain can be worked out. This can be useful if one chain is damaged as it can be repaired using the information on the other chain. We can imagine a hypothetical piece of a DNA chain as follows, where the letters represent the bases:

— A T C A G C T T C A G ——
— T A G T C G A A G T C ——

The double-stranded helix immediately suggested how the genetic material makes copies of itself. To replicate, the two strands unzip, assisted by special unwinding enzymes, with new chains produced by base pairing according to a fixed rule. In this way, each strand serves as a "template" to form a new sister strand from a pool of free bases floating about nearby in the nucleus. To ensure that the copying is flawless, remarkable enzymes snip out any wrong base pair and correct the mistake. The result is the creation of two identical DNA molecules. The replication of the DNA molecule is illustrated in Figure 5.

This mode of replication is semi-conservative in that one strand of the original DNA molecule is conserved intact in each of the two new molecules. This ensures that the newly formed DNA molecules are identical to the parental molecule, an important prerequisite for a molecule responsible for the genetic continuity of living organisms.

In addition to understanding how the DNA molecule makes copies of itself, the other important issue was, how does nuclear DNA influence the rest of the cell. The DNA molecule contains all the necessary information that makes a cell work.

Figure 5
DNA replication

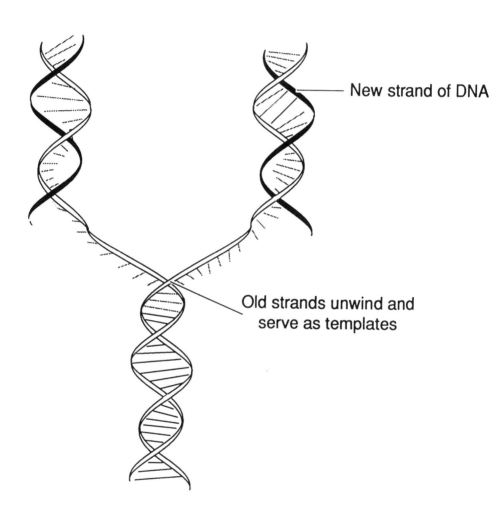

New strand of DNA

Old strands unwind and
serve as templates

It determines everything from the colour of our eyes to the detailed chemical reactions of digesting a morsel of food. It does this by controlling the synthesis of proteins, which are at the hub of the machinery of life.

The base sequence of the DNA molecule carries the information that decides the sequence of amino acids, the little units from which proteins are made. This presented a few problems to biologists. First, as DNA is in the nucleus of the cell and protein synthesis takes place outside the nucleus in the cytoplasm, there must be a means of sending the message from the nucleus out into the extranuclear provinces. This messenger is another molecule of nucleic acid called RNA for ribonucleic acid. Each messenger RNA is made from a stretch of DNA by simple base pairing, and passes to the cytoplasm where it controls the construction of one protein molecule.

This raised the problem of the genetic code. Somehow, the four different kinds of bases in DNA and RNA must specify the twenty amino acids found in proteins. Biologists had correctly speculated that groups of three adjacent bases coded for individual amino acids. Groups of two bases, such as AT or TC, could be arranged in only 16 different permutations (4 x 4 = 16), too few to accommodate all twenty amino acids. However, groups of three bases, such as AGT or GAC, yield 64 possible permutations (4 x 4 x 4 = 64). Sydney Brenner and Crick proved that the four bases of the DNA molecule specify the twenty amino acids found in proteins by means of a triplet code or codon. It turns out that all 64 codons are used. In the standard code, two amino acids have only one codon each, many have two, one has three, several have four, and two have six codons. Three other codons, not used to specify amino acids, are signals for stopping the synthesis of the protein chain.

What Brenner and Crick also demonstrated was that the sequence and number of amino acids was all the information required for protein synthesis. The long, string-like molecules fold by themselves into specific three-dimensional structures, which determine the properties of each protein. Before this, it was widely held that proteins had no definite structure, and the relevant issue was to find out how the amino acids were joined together and what provided the energy to hold them together. Some had suggested that the genes controlled the energy flow for protein synthesis. In reality, however, the role of DNA is simpler and more direct; it contains the blueprint for the arrangement of the amino acids in the protein molecule.

In 1961, Marshall Nirenberg performed a series of experiments that deciphered the genetic code. He found that a synthetic RNA molecule that contained the base uracil (U) — the RNA equivalent of thymine — coded for a protein made up of a single amino acid; in this case phenylalanine. From this, he concluded that the codon UUU coded for phenylalanine. Similarly, other triplet codons were identified, each coding for a specific amino acid.

The genetic code, the dictionary that relates the four-letter language of nucleic acids to the twenty-letter language of proteins, is used by all living things on earth. The uniformity of life is astonishing. We share genes, and the resemblance of the enzymes of the yeast to those of humans is remarkable. Living creatures carry stores of information in their nuclei that are more striking in their similarity than their diversity. They obey the same rules of replication and drive the cells' machinery using the same biological manual. There are some minor variants in some of the micro-organisms but, in the end, the basic rules still appear to be inflexible.

The discovery of the role of DNA in directing the activities of the cell, which is what metabolism is about, by synthesizing messenger RNA, each of which, in turn, controls the construction of one protein, had a major influence in the 1950s in focusing attention towards the flow of information in the cell. A central hypothesis was put forward to depict this:

DNA —> RNA —> PROTEIN

This appears to be the common state of affairs, but important exceptions are to be found in nature. In certain viruses that contain an RNA molecule as their genetic material, the transfer from RNA to RNA is used. Another special group of RNA viruses, the retroviruses, have the unique ability to transfer information from RNA to DNA. The transfer from protein to DNA does not appear to occur.

CHAPTER 3

LESSONS AT THE THRESHOLD OF LIFE

The existence of viruses was suspected from mediaeval times, although the term was originally used in a somewhat loose sense to denote all sorts of poisons and noxious substances. By the mid-1800s, a more systematic approach to understand the causes of disease led to the recognition that there are distinct types of harmful agents. The chemical poisons, such as carbon monoxide or potassium cyanide, were described, but it soon became evident that a lot of illnesses could not be explained.

A major milestone was reached in 1857 with the brilliant work of a French chemist at the University of Lille. Louis Pasteur was interested in what caused wine and milk to go sour. He showed that fermentation of these liquids was due to a germ, and the type of fermentation depended on the type of micro-organism. He was able to prevent the process by boiling the liquid, and then sealing it off from air. From this, he concluded that the putrid fermentation was caused by microscopic airborne germs, and announced this to the world in 1864.

The following year, a Scottish surgeon, Joseph Lister, the son of Joseph Jackson Lister who developed the achromatic microscope, learned of Pasteur's work. He immediately implemented the practice of antisepsis in the operating theatre in an effort to prevent suppuration of surgical wounds, then a serious problem in the care of patients. Lister thought that bacteria might be responsible for wound infections, and his technique of washing open wounds and spraying the air of the operating rooms with carbolic acid, and applying protective dressings to keep bacteria from contaminating wounds substantially reduced post-operative infections. He got the idea of using carbolic acid from the observation that adding carbolic acid to the town sewage during an epidemic in cattle at Carlisle, helped the cows recovered.

Rapid discoveries followed with the identification of some of the major bacterial killers like the tubercle bacillus. Doctors soon realized, however, that chemical poisons or bacteria could not account for all diseases. There appeared to be a different class of agents which, like poisons, could pass through filters sufficiently fine to hold back the smallest bacteria, but, like bacteria, could multiply in affected persons to cause illness. But it was not possible to grow them in the laboratory with the methods used for bacteria. Even when provided with nutrients known to be required by the most fastidious bacteria they did not grow. Although not visible under the ordinary light microscope or cultivable under routine laboratory conditions, they were recognized as the cause of such deadly diseases as smallpox and rabies. To this group of "infinitesimally small" microorganisms, as Pasteur called them, the name filterable viruses, or later, simply viruses, was given.

In 1892, a Russian botanist, Dmitri Ivanowski, studied the mosaic disease of the tobacco plant, and found that sap from diseased plants transmitted the disease to other healthy plants, even after it was filtered to remove bacteria. This was the first demonstration of the infectivity of a virus. Ivanowski's work was corroborated six years later by a Dutch botanist, Martinus Beijerinck. In 1898, G. Sanarelli, an Italian working in Uruguay, discovered that a tumorous condition, multiple myxomatosis, in rabbits was caused by a virus. Over the next few decades, biologists confirmed that viruses can cause a variety of diseases in both plants and animals, and also became aware of yet another group of viruses, the bacteriophages, which are specially adapted to infect bacteria.

The biological science of recent years has brought us some truly amazing new information concerning the simplest organisms. The viruses exist at the threshold of life, capable of replication and containing pieces of genetic material that can code for proteins. Yet, they are inert until they infect a cell, which is then held hostage and forced to use its own machinery to make more viruses. Certain viruses have become masters at the biological game of life, and snatch vital pieces of DNA from the cell they infect and add these to their own genetic material. The discovery of this family of viruses has raised some troubling concerns because a few of its members are implicated in some cancers. Viruses can be the agents of diseases, from the common cold to the uncommon encephalitis; and, as we now know, they can also cause cancer.

Although the effects of the virus could easily be observed, their identification was to remain elusive for a long time. Viruses are extremely small. Their size is such that they are most conveniently measured in the units of the micron, one-thousandth of a millimetre, and a millimicron. For comparison, the blood lymphocyte, one of the smallest mammalian cells, is about 10 microns in diameter. This is several times larger than typical bacterial cells, which are of the order of 2 to 4 microns. Viruses are much smaller, running from about 20 millimicrons to about 250 millimicrons. The smallest viruses range down to the dimensions of large molecules; the egg albumin molecule, for example, is 2.5 by 10 millimicrons.

By 1935, Wendell Stanley identified the virus responsible for the tobacco mosaic disease. This was a laborious task, as he managed to extract the virus from a ton of infected leaves. But his efforts prompted an intensive study of the nature of viruses. In store, were many surprises.

The simple structure of viruses is their distinctive feature. They are minute particles made up largely of an outer protein shell, within which is a small amount of nucleic acid, the obligatory molecule for membership in the club we call life. The viral genetic material consists of strands of nucleic acids, either DNA or RNA. The amount of genetic information in a virus ranges from 3 to 300 kilobases. Since an average gene is about one kilobase, the number of genes in a virus varies from one to a few hundred, a small number compared with other living things.

In 1952, Alfred Hershey and Martha Chase made an important discovery. They found that when a bacteriophage infects a bacterium, almost only its DNA and very little of the bacteriophage's protein enters the bacterial cell. By then, Avery and McCarty had shown that pure DNA was enough for transmitting genetic information from one bacterial strain to another. It was not difficult to take this one step further and ascribe infectivity of the bacteriophage to its DNA. The protein shell serves merely as a vehicle for transporting the nucleic acid strands. Viruses were beginning to look like mobile genes, darting around from organism to organism.

A problem arose when the RNA virus, such as the tobacco mosaic virus, was discovered. At that time, no precise genetic role was known for the simpler RNA molecule, and many virologists thought that the protein of the RNA virus was its "genetic" material. Working with the tobacco mosaic virus, Heinz Fraenkel-Conrat in California and Gerhard Schramm in Germany successfully extracted its RNA, and were able to transmit the mosaic disease by applying it to the leaves of uninfected tobacco plants. This demonstration of the genetic potential of viral RNA, proclaimed in 1956, has been established by subsequent experiments, which confirmed that RNAs from plant and animal viruses carry the information necessary to produce diseases. The RNA molecule ascended to a new position in the genetic hierarchy.

The understanding of how a virus infects a cell and what then happens called for thinking along new lines. Unlike living cells, viruses do not grow in size and divide because they lack the cellular infrastructure to support cell division. Instead, they are strict parasites. A relatively small piece of DNA or RNA packed into a protein shell can invade a cell, which is then forced into new, alien metabolic pathways, and whose machinery is harnessed into producing viral parts instead of cellular parts. While the energy and raw materials are those of the infected cell, the blueprints for the finished product are supplied by the genes of the invading virus. In the final step, the nucleic acids become wrapped by viral protein coats, and a progeny of new viruses burst out of the cell. These may invade the nearby cells, or be transmitted through tiny droplets in a sneeze to infect another person.

The simple structure of the virus with its small piece of nucleic acid may, at first glance, suggest that its replication in the infected cell is a cinch. For the DNA virus, the basic plan is simple enough. To begin with, the two chains unzip and synthesis of the new chains proceeds. There are special proteins whose job it is to unwind the double helix, allow the chains to rotate around each other, and then join up again. When all is done, two identical DNA molecules are produced. The rules for this process are precisely the same as for replication of the DNA in a human cell. Size is no excuse for short-cuts.

The replication of an RNA virus presents some special problems. We will recall that the usual flow of genetic information in most living things is:

$$DNA \longrightarrow RNA \longrightarrow PROTEIN$$

The DNA virus observes this pattern. However, in the RNA viruses, the RNA commands a supreme position as the genetic material and multiplication of an RNA virus must, therefore, require some novel mechanisms. The viral RNA serves as a template for the synthesis of new viral RNA, possible because of special viral enzymes. What is unusual about this is that the entire process is primed by the RNA itself, and a DNA molecule is not involved at any stage.

This is, however, not the sole means of replication of the RNA virus. In 1970, Howard Temin and Satoshi Mizutani discovered another class of RNA viruses, called retroviruses. The way in which the retrovirus replicates is extraordinary, and called for a modification of the central dogma of the flow of genetic information. The retroviral RNA is transcribed into a DNA by a unique enzyme, reverse transcriptase. This was a startling discovery. It was the first demonstration of the flow of genetic information from RNA to DNA. Even in the typical RNA viruses without reverse transcriptase, where the RNA serves as a template for duplication of the RNA genetic sequence, such a reversal of genetic flow with a DNA intermediate is not observed.

The retroviruses proved to be of considerable biological significance. They are a large family found in a wide variety of animals and man. Two human retroviruses, HTLV-I and HTLV-2, cause different types of leukaemia in adults. A third retrovirus, HIV, is of major interest as the virus responsible for AIDS.

The idea that viruses play at least a part in causing some types of cancer is not new, although it became generally accepted only recently. The notion that cancer might be caused by a virus goes back to 1908 when Vilhem Ellerman and Oluf Bang speculated that leukaemia in chickens seemed to be caused by a virus. They successfully transmitted the virus from fowl to fowl, producing leukaemia each time. In 1910, Francis Peyton Rous described a particular cancer called sarcoma, also in chickens, that was due to a virus now referred to as the Rous sarcoma virus. Rous had demonstrated the virus by its biological effect; he did not have the means to directly observe it. He took cancer tissue, ground it up, and injected a portion into healthy chickens. These chickens then developed cancers. But the suggestion that a virus caused cancer was so controversial and radical at the time that Rous did not use the word to describe the agent responsible for the tumour.

Half a century later, in 1966, Rous was awarded the Nobel

Prize for Medicine and Physiology for his discovery of tumour viruses. It was the longest time lapse ever between a discovery and the award of the prize in the history of the Nobel Prize committee. Why had it taken so long? He had considerable difficulty in convincing the scientific community of the importance of his finding. A human cancer is not a transmissible disease; you do not get cancer from your neighbour in the same way you catch a common cold or the flu.

The scepticism that surrounded much of the early work of the viral origin of cancer arose largely through the inability to observe the virus directly. Seeing is believing. But viruses are too small to be seen under a conventional microscope, which examines objects by means of a light, and whose limit of visibility is about one-fifth of a millimetre. Not until the availability of the electron microscope, which uses a beam of electrons instead of light, could much smaller objects be seen.

The first virus to be associated with a human cancer was discovered by Anthony Epstein and his assistant, Yvonne Barr, and is now called the Epstein-Barr virus. The virus is linked to Burkitt's lymphoma, named after Denis Burkitt, a British surgeon, who first described it. Burkitt suspected that an infective agent could be responsible for the malignant lymphoma that struck children in East Africa. The clue came from the geographic distribution of the cancer, which seemed to depend on temperature and rainfall. Burkitt described his observations at a lecture at the Middlesex Hospital Medical School in 1961, and this led to a fruitful collaboration with Epstein, who realized that the influence of climate on the occurrence of the cancer could be related to a virus. The Epstein-Barr virus turns out to be a DNA virus belonging to the herpes family, whose other members cause cold sores, chicken pox and shingles.

But it is from the study of the RNA retrovirus that a wealth of information about the origin of human cancer has been gained. When an RNA retrovirus infects a normal cell, an intermediate double-stranded DNA molecule, called a provirus, is synthesized from the viral RNA molecule under the direction of a special viral enzyme, reverse transcriptase. This newly formed DNA then becomes integrated into the cellular chromosomal DNA, where it enjoys all the privileges of its host DNA. It is replicated along with the cellular DNA in dividing cells, and its message, coded in its particular sequence of bases, is translated using the host cell's enzymes.

The provirus has violated the myth of the diversity and separateness of living things. The unity of living creatures is far greater than was supposed at the turn of the century. When it comes down to it, we share with all life a fantastic and elaborate single chemical language. At the very heart of life, the nucleic acids that carry genetic information are identical in all the micro-organisms, plants, animals, and man. The same molecules are used over and over in varying patterns, conservatively and ingeniously for different functions.

The key to unravelling how a retrovirus causes cancer was the realization that some members of the family, such as the Rous sarcoma virus, very quickly induce tumours in infected animals. In contrast, others appear much less efficient. These weak viruses cause cancers only after long periods, usually several months, in infected animals. The prototypic example of this class of viruses is the avian leukosis virus, which causes lymphomas in chickens. Weak retroviruses of this type are commonly found in nature.

The difference in the ability of the two types of retrovirus to cause cancers carried a critical implication. It appeared pos-

sible that the Rous sarcoma virus contained genetic information that was absent in the avian leukosis virus. In 1970, Peter
Duesberg and Peter Vogt found that the RNA of the Rous sarcoma virus was longer than that of the avian leukosis virus. The
obvious suggestion was that the extra piece of RNA contained
the information responsible for the special ability of the Rous
sarcoma virus to induce cancer. This was a remarkable discovery. The tumour virus carries in its nucleic acids a gene, which
it brings into an infected cell, where its expression leads to a
cancer.

The typical retrovirus in nature, such as the avian leukosis
virus, contains three coding segments in its genetic RNA. The
gag gene encodes core proteins which are the major structural
proteins of the virus. The *pol* gene encodes the important
enzyme, reverse transcriptase, for synthesis and integration of
the proviral DNA. The *env* gene encodes the envelope proteins
on the surface of the viral particle.

These three genes are all that the typical retrovirus found
in nature needs to create a progeny of new retroviruses. Once
it invades the cell, the enzyme of the *pol* gene, reverse transcriptase, which the virus carries with it, sets about to camouflage
the viral RNA as a double-stranded DNA, which becomes
securely lodged in the cell's chromosomal DNA. It is a clever
ploy. From this secured position, the viral genetic material can
usurp the plentiful cellular resources and raw material in its
single-minded drive to produce more viruses. In the final stage,
the new copies of RNA are packed into viral particles. The host
cell is sacrificed, and a new generation of viruses is released. A
few thousand bases, strung together to form the viral RNA,
know extraordinarily much; they know how to insinuate themselves into an innocent cell and make copies of themselves at

the host's expense.

If the replication of the ordinary retrovirus can be so efficiently controlled by its three genes, then the extra genetic material in the virulent tumour virus must be responsible for its proclivity to induce cancer in susceptible animals. When the genetic material of the Rous sarcoma virus was examined, a fourth gene was discovered. This is the cancer gene or oncogene. Because the Rous sarcoma virus causes sarcoma in chickens, its oncogene, the first such gene to be identified, was called the *src* oncogene. All of the other virulent tumour viruses, like the Rous sarcoma virus, contain at least one viral oncogene. In some cases, the same oncogene is found in different viruses. There are over two dozen different oncogenes now described in the genetic RNA of about forty tumour viruses in several animals.

The presence of oncogenes in the viral RNA seems paradoxical because they do not play a direct role in the replicative life cycle of the virus. But genes are not created from nothing; they must come from somewhere, genes evolve. An inquiry into the origin of the viral oncogenes and how they ended up in the virus uncovered some unexpected surprises. It turns out that the retrovirus picked up its oncogene from the chromosome of a normal cell at a earlier time in evolution. Once this gene is grafted into the viral genome, it is tugged along from one animal to another.

In 1976, Harold Varmus and Michael Bishop provided the first direct evidence of the origin of the retroviral oncogene. They searched for the *src* gene in healthy chickens. The result was startling. The chromosomes of normal chicken cells contain a *src* gene. The very gene that induces cancer when delivered into a healthy cell by a retrovirus is normally found in that cell. This was taken even further. The *src* gene was looked for in several other animals, as well as humans, and in every case, a similar gene was found. This established that normal cells contained genetic information closely related to a viral oncogene. It was beginning to look as though all cells normally carry genes that might have the potential for causing cancer — a disquieting prospect. In recognition of their landmark studies, Bishop and Varmus were awarded the 1989 Nobel Prize in Medicine and Physiology.

The discovery of the *src* oncogene in normal cells prompted a search for the other known viral oncogenes. In every instance, similar genes that bore a striking resemblance to those in the retroviruses, were found in normal cells. It now appears that each of the retroviral oncogenes has originated from the genes of normal cells. The normal cellular genes from which the viral oncogenes are derived are called proto-oncogenes.

The presence of proto-oncogenes in the chromosomes of normal cells at once raises the question of their possible function. The implication of this is that since the related viral oncogenes cause cancer in vertebrate animals, do proto-oncogenes contribute to the formation of cancer in humans? This is obviously not an issue to be taken lightly. We carry stores of DNA in our nuclei, pieces of which resemble viral oncogenes that induce cancer, and may have, in fact, at one time been copied and grafted into the viral genome from us. If this is true, and the evidence is compellingly in favour of this view, then the retroviruses are more skillful at the biological game than it seemed just a decade ago. These simplest of life forms gain access to the deep interior of our cells, where they replicate in their own fashion, privately, with their own genetic information, and before taking their leave, swipe the information encoded in some piece of the chromosomal DNA.

Some important lessons were learned with an understanding of how the viral oncogene goes about the business of causing cancer. The first obvious explanation is that once a proto-oncogene becomes transplanted into a retrovirus, it becomes part of the virus. As such, the oncogene is under the control of viral switches, unlike its normal counterpart that is precisely regulated. When the viral genome, along with its oncogene, is inserted into the chromosomal DNA, there is a frenzied translation of its information into protein molecules. The consequence is an overabundance of the oncogenic protein molecules. If this protein plays a pivotal role in the regulation of cell growth, unbridled cell proliferation could occur.

An alternative explanation is the mutational hypothesis. When a proto-oncogene is picked up by the retrovirus, a small but critical change or mutation in the base sequence occurs. This mutation in the oncogene eventually shows up in the protein for which it codes. As the function of any protein depends on its three-dimensional configuration, even small changes in the sequence of its amino acids can cause major distortions, and the protein malfunctions.

The retroviral oncogenes come from normal cellular genes, the proto-oncogenes, which are widely conserved throughout evolution. The proto-oncogenes can be found in the cells of unrelated species, from humans and animals to lower organisms, such as the fruitfly and even yeasts. This argues for a common evolutionary kinship.

The proto-oncogenes are vital for normal cell growth and development. The viral oncogenes differ from their normal progenitors, and through an overabundance or structural alteration, are capable of inducing a cancer in a susceptible animal. The discovery of the viral oncogene paved the way to identifying the genetic abnormalities that contribute to the formation of human cancer. The evidence that we harbour genes with a potential to cause malignancy is strong, and direct.

CHAPTER 4

BETRAYAL FROM WITHIN

I n the early 1900s, Theodor Boveri put forward the hypothesis that a cancer arises from a normal cell as a result of a derangement of its chromosomes. His idea was the culmination of a slow progress in understanding the nature of cancer that can be traced back to the work of Müller in the mid-1840s.

Not that progress was not made on other fronts. Pathologists carried on the tradition of Bichat with precise descriptions of more than two hundred different kinds of human cancers, each named for the cell type of origin; for example, melanomas, cancers of the pigment-producing melanocytes that are densely located in moles; or hepatomas, cancers of liver cells or hepatocytes. Most of these were divided into three major groups: carcinomas, sarcomas, and leukaemias-lymphomas. Carcinomas, which make up nearly 90 per cent of human cancers, arise from cells of the sheets or epithelia that cover our surfaces, such as skin or gut. Sarcomas include cancers of the supporting tissues, like bone, muscle, or blood vessels. Leukaemias and lymphomas develop from circulating cells of the blood and lymph systems. The sarcomas and the leukaemias and lymphomas make up about 10 per cent of human cancers.

Cancers were also classified according to the tissue in which they arise, for instance, lung, breast, or colon cancer, and by their microscopic appearance. A lung cancer could be of different cell types — squamous cell carcinoma, adenocarcinoma, large cell carcinoma, or small cell carcinoma. As more and more data were kept on the various cancers, doctors recognized that they differ in how they grow and spread. They were able to predict the outcome for each type of cancer, and how therapy alters the natural course of the disease.

Epidemiologists used the science of statistics to study the distribution of cancer in the population, and plot the rise and fall in the frequency over time. An extensive data base was prepared. More than just accumulate every new bit of information, this enterprise has allowed physicians to come up with some valuable advice in preventing or avoiding certain cancers. The most outstanding example is the role of cigarette smoke as the cause of lung cancer, the most lethal cancer in the western world.

Much of what is done in the treatment of cancer, by surgery, radiation therapy, and chemotherapy, have been empirical, in the sense that they are employed to remove or destroy the malignant tissues, regardless of the mechanism by which the cell becomes malignant. The problem has been that little was known about the process of the formation of a cancer. But much has happened in the past two decades to enable scientists to lay out in sharper detail the changes that cause a cell to go from normal to malignant. The retrovirus has been the key that opened the door to this new insight. The first suggestion of a tumour virus dates back to 1910 when Rous speculated that sarcoma in chickens might be due to a virus. The significance of this was not immediately realized, but the tremendous advances in modern biology have allowed scientists to assemble a biological Rosetta stone with the tumour virus providing the code to decipher its message.

In 1967, Robert Huebner and George Todaro proposed the "oncogene" hypothesis. On the basis of the information gathered since the studies of Rous, they postulated that animal cells contain genes, the proto-oncogenes, which are not normally active, but if they are switched on, could induce the development of cancer. They also speculated that the genes of the tumour virus could trigger the proto-oncogene from a quiescent state to an activated one. At the time, this was a bold hypothesis as it squarely puts the cause of the problem in the heart of the cell, its nuclear genes.

The discovery that the RNA tumour virus or retrovirus, which causes cancer in animals, contains two types of genes — those that regulate replication of the virus, and those that incite cancer — was the first clue of the existence of cancer genes. The next piece of evidence was finding oncogenes in the cancer cells of animals. This in itself was not sufficient proof, as it was quite possible that these genes were pieces of the viral genetic material that strayed into the nucleus. Nonetheless, their presence was enough to stir an intensive investigation that is still going on in earnest.

A radical change in thinking about the origin of human cancer occurred in the 1980s. By then, cellular proto-oncogenes, which bear a striking and undeniable resemblance to the viral oncogenes, were found in normal cells from a wide range of animals, including humans. It seemed that wherever a careful search was made, a proto-oncogene was uncovered.

A remarkable story was slowly pieced together. The retroviral oncogene that induces cancer in animals originates from the proto-oncogene of animals and humans. When it is taken up by the virus, it is modified in a small but critical way, yet not so much that the family resemblance is lost. In a healthy cell, the viral oncogene exerts a devastating effect. Outside the bounds of the cell's normal regulatory devices, the viral oncogene provokes rampant cell division. A cancer comes into being.

At the turn of the century, Boveri, with a prophetic vision, had correctly predicted that the answer to the problem of the genesis of cancer is to be found in the nuclear chromosomes. No one seemed moved by the idea then, but thanks to the new science of molecular biology, we now have an intimate understanding of the cell. With this comes compelling evidence implicating the cellular genes in cancer.

The first proto-oncogene known to exist inside animal and human cells was discovered in 1976 by Michael Bishop and Harold Varmus at the University of California in San Francisco. Since then, about fifty proto-oncogenes have been identified. When we consider that the typical human cell contains a pool of over 130,000 genes, the proto-oncogenes are few in number. Quite possibly, more proto-oncogenes might be found, but it now seems the total number that are important in human cancer is small.

Normal cells from distantly related species contain proto-oncogenes. The ancestors of these genes have been passed on and conserved through evolution, undergoing slight changes over the vastness of time, but their fingerprints leave no doubt of their identity. Nature has used the same mould to create the wondrous diversity of new species through the resculpturing of a few critical genes.

It might be evident by now that the term proto-oncogene, connoting as it does a relation to cancer, is a misnomer for this special family of genes. This merely an accident of their discovery in the tumour virus. In reality, the proto-oncogenes are the regulatory elements in the growth and development of living organisms.

Just how important the proto-oncogenes are can be easily gleaned when we consider the exquisitely co-ordinated series of events that go into the making of a human. There is no obvious resemblance between the nearly spherical fertilized egg and a baby at birth or an adult. How does this single cell develop into an independent person with eyes to see the world, a brain to think, or ears to enjoy a concert?

Classical embryologists held that the egg and the sperm are not entirely shapeless, but must contain a miniaturized version of an adult. In the 1700s, this prevailing notion was taken to its extreme with the theory of preformation. The egg and the sperm were thought to be actually tiny adults, with the function of development reduced simply to making this hidden little person bigger.

We now know that the mature adult develops from the single fertilized egg. The challenge was to find out how the immense variety of cell types, the nerve cells, the muscle cells, the blood cells, and so on, are spawned from a single cell. This specialization or differentiation of cells begins soon after the egg starts to divide. The result is the transformation of the original cell into a large number of different daughter cells, each specialized to fulfil a certain function.

The complete instructions for development are written in the linear sequence of bases in the DNA of the fertilized egg. In a fantastically co-ordinated manner, the parental egg sets the schedule of which segments of the DNA are read. Groups of unique proteins determine whether the cell is a thyroid cell secreting its thyroid hormone, or a muscle cell of the heart contracting and relaxing as it pumps blood through the body. Cells communicate with each other by simple touching, or secreting hormones that are sent off to react with other cells. All of these activities ultimately come from the orderly, regulated expression of the DNA.

Many hundreds of proteins are responsible for making a muscle cell what it is, and a different group of as many proteins are needed by a liver cell. Something must tell muscle cells to manufacture the various proteins used to construct muscle fibres, and, at the same time, not make those proteins for digesting food. This means that only certain genes are active in the differentiated cell. An explanation that immediately comes to mind is that only a fraction of the genes in the fertilized egg are passed on to each cell type. But this is not the case. The cells of the body contain the same chromosomal complement; all cell divisions are preceded by a regular mitosis, so that the daughter cells receive an identical set of chromosomes. There is no loss or gain of chromosomes. Instead, it appears that something has happened to have turned off the cell's capacity to synthesize proteins other than those required for its particular task. Differentiation is controlled by devices that selectively throw the switch on specific genes; cells are wired very early in development to specialize.

Many differentiated cells are destined never to divide again. The red blood cells are a prime example of this. They manufacture the protein, haemoglobin, that carries oxygen in the blood, and the entire cellular machinery is set up for its production. After an average life span of 120 days, the red cells wear out and are destroyed in the spleen. As they are removed from circulation, more red cells are produced in the bone marrow to replace them. In contrast, other specialized cells, such as nerve cells, though never dividing, live many years.

Traditionally, one of the problems that impeded an understanding of how the flow of information from the chromosome is regulated in differentiation, is the incredibly large number of genes involved. Many hundreds of specific protein molecules go into the making of a nerve cell, requiring at the very least as many genes. To study the details of the function of these is a daunting task. It turns out, however, that genes can be arranged in a hierarchy, with some genes controlling the expression of other genes. The genes of differentiation are not equal. Some are more important than others, because they control the decisions to specialize along one pathway or another. The proto-oncogenes play important roles in the proliferation and differentiation of cells from the early embryo and well into adulthood, where many cells are still being produced to replace dying cells. The proto-oncogenes are at the top of the heap.

The earliest stages of embryonic development are controlled by maternal RNAs that are synthesized and stored, often for long periods, in the egg. After fertilization of the egg by a sperm takes place, the egg then begins a series of cell divisions. This is jump-started by the maternal RNAs, but after just a few rounds of cell division, further development is fuelled by several of the proto-oncogenes. They are indispensable to normal growth and development.

The retrovirus brought us the oncogene and changed the course of biology. New information about the way DNA works is being gathered at an ever-increasing pace. Curiously, all this is taken calmly, without much upheaval. The news about the discovery of DNA and the genetic code in the 1950s was greeted with enthusiasm, but now we seem to take much for granted as if every piece of new knowledge just fits nicely into an empty niche. Perhaps, it is precisely because the new data do not overturn or displace any prior dogma that we have come to accept them so readily.

From the retroviral oncogenes came the discovery of activated oncogenes in human cancers. This has been a major stride in the quest to understand the disease and, as if this is not enough, we were soon to learn that these very genes are central to the evolution of life. We got more than we bargained for.

When things are going well, the proto-oncogenes orchestrate the growth and development of the organism. They become active at certain times, and once they have performed their piece, become quiet until called upon again. But the mechanism is not infallible. Slight alterations in the proto-oncogene disturb the programme of orderly growth.

How does this happen? The answer to this question is obviously of paramount importance. It called for a new direction in thinking about the problem. In the past, biologists wrestling with the puzzle over what makes a cell become malignant, have looked at external forces that act on it. They have considered such inciting causes as chemical carcinogens, environmental toxins, radiation exposure, or any of a multitude of other causes. Although they knew that the DNA was damaged in some way by these outside influences, the trouble was that there was no way of probing the deep interior of the cell. The modern technology of molecular biology makes this possible today, so matter-of-fact that it is almost routine. And what we have learned is that the target is precise, intimately related to the very essence of the life of the cell. Cancer is, in the end, a betrayal from within.

With over fifty proto-oncogenes, the task of finding out how each works seems difficult at first, but this became much simpler when it was realized that, broadly speaking, they act in one of two ways. Either the oncogene is switched on at the wrong time during the life of the cell, or there is a mutation within the oncogene.

For some time, biologists had repeatedly observed that some types of human cancers had certain chromosomal abnormalities. A piece of one chromosome may break off and become attached to another. In these circumstances, a group of genes is moved to a new chromosome, where they come under the influence of a different regulatory switch. If it so happens that the switch is in the on position, the transported genes become accidentally activated. Without an in-depth knowledge of the exact make-up of each gene, this reshuffling of genetic material in cancer cells remained little more than a laboratory curiosity.

It is now possible to map precisely the location of the known proto-oncogenes in any of the twenty-three pairs of chromosomes in the nucleus. It did not take long for this information to be merged with the description of the chromosomal defects. The trick was to find out if the proto-oncogenes are carried in the fragments of chromosomes that are displaced.

A revealing example of the consequences of chromosomal rearrangement is seen in Burkitt's lymphoma. A frequent finding in this disease is that a fragment of chromosome number 8 containing a proto-oncogene, the *myc* proto-oncogene, moves to chromosome 14, whose genes code for antibodies. Because the malignant cells of Burkitt's lymphoma are derived from cells that continuously produce antibodies, an important clue was provided. By leaving its native chromosome, the *myc* proto-oncogene is deprived of its regulatory switches, and its relocation on chromosome 14 puts it in a foreign province, an active antibody-coding region. The *myc* proto-oncogene, bowing to a new authority, allows free access to its coded information. As its message gets around, the cell's everyday function is subverted, and it responds to the provocation by countless cell divisions.

The frequent relocation of the *myc* proto-oncogene into an active or "hot" region of a chromosome in Burkitt's lymphoma suggests that this is an important means of proto-oncogene activation in human cancers. Such, indeed, is the case. Another example of this is observed in a type of leukaemia called chronic myelogenous leukaemia, in which there is the telltale chromosomal abnormality, recognized as the Philadelphia chromosome.

In 1958, Peter Nowell and David Hungerford reported a distinct anomaly in the blood cells of two persons suffering from chronic myelogenous leukaemia. Named after the city in which it was discovered, the Philadelphia chromosome is the hallmark of this form of leukaemia. It is present only in the leukaemic cells and not in normal tissues. When the disease is in remission, the chromosomal abnormality also disappears. The Philadelphia chromosome results from an exchange of chromosomal fragments between chromosomes 9 and 22. Chromosome 22 is one of the smallest chromosomes, a fortuitous circumstance, because although it loses nearly half of its chromatin content, this represents only a small fraction of the total chromatin of the nucleus. Such a small deletion might not have been discovered so early had it occurred in one of the larger chromosomes.

The *abl* proto-oncogene is on chromosome 9, and the *sis* proto-oncogene on chromosome 22. In chronic myelogenous leukaemia, the movement of the *abl* proto-oncogene from chromosome 9 to 22 is the critical event. This places it in an active antibody-coding region on chromosome 22, again leading to its untimely and inappropriate activation.

In addition to the untimely activation of a proto-oncogene, some undergo changes or mutations that set the cell on the track to becoming malignant. A mutation in the DNA causes a change in the amino acid sequence of the protein for which it codes. As a consequence, the three-dimensional configuration of the protein molecule is altered and it malfunctions.

The *ras* proto-oncogene provides a good example of the dire effects of a genetic mutation. In 1982, Robert Weinberg, Mariano Barbacid and Michael Wigner, working independently, found that a single mutation converted the normal *ras* proto-oncogene into a powerful oncogene. In some cancers, a single base in the proto-oncogene is changed in a codon from GGC to GTC — a point mutation. The codon GGC codes for the amino acid glycine, whilst GTC codes for valine. When the sequence of bases of the *ras* proto-oncogene is read into a protein molecule, an error occurs with the substitution of the amino acid valine for glycine. At first glance, such a minor change may seem inconsequential, but within each protein there are a few critical sites upon which the proper folding of the amino acid chain depends. Tampering with these causes a distortion of the three-dimensional structure of the molecule and interferes with its normal function. The change of a single amino acid, at just the right spot, can be devastating.

What is remarkable about this point mutation in the *ras* proto-oncogene is that it represents a change in a single nucleic acid in a gene of over 5,000. Except for the point mutation, the mutant *ras* gene is indistinguishable from the normal proto-oncogene. The discovery of the serious consequence of a point mutation in the *ras* proto-oncogene was of exceptional significance. For the first time, a human cancer can be traced to a precise genetic lesion.

Proto-oncogenes are indispensable to normal growth, but their untimely expression or mis-expression due to a mutation wreaks havoc. To understand how this is brought about, we must examine the events that follow the activation of the proto-oncogene. The issue at hand here is: What are the proteins coded by these genes? Such an approach is fruitful because it is left ultimately to the proteins to carry out the mission of the genes.

The protein of the *src* oncogene of the Rous sarcoma virus that causes cancer in chickens is of historic importance, as it was the first to be studied in detail. Defining its structure was straightforward enough, but elucidating its function was not as easy. It is an enzyme of a small group that regulates other cellular proteins, but seemed, at the time, to be unique with respect to the proteins it affected. Not long after, however, this was shown not to be the case. Protein molecules that could be targets of the *src* enzyme were found in normal human cells. This meant that an enzyme of the same ilk as *src* must, therefore, be present in human cells. A vigorous search ensued, from which came a startling discovery. An enzyme was, indeed, found in normal cells, and it turned out to be none other than a human version of the viral *src* enzyme. This was another in the chain of endless discoveries that revealed how blurred the line between normalcy and malignancy was becoming.

As the proteins coded by the proto-oncogenes were examined, a common theme emerged. They form an elaborate network that links the millions upon millions of cells in the body together. Without a close integration and co-ordination, there would be chaotic growth. Cells are interdependent, inseparable, all working for the common good. The proto-oncogenic proteins make up the circuits for cellular communication. This is achieved by special chemical molecules or growth factors, which are embraced by matching proteins or receptors on the surface of the cell. And while they dance, their whispers are passed on to the interior of the cell by a complex of proteins, the G proteins. Within the cell, another group of small molecules, the chemical messengers, relay the information to the internal switchboard, the nucleus, which responds by issuing its own precise and unequivocal instructions to the cell about all kinds of matters on when to divide and how to differentiate.

It is an intricate and wonderful mechanism for the exchange of messages between cells. The social chatter goes on ceaselessly, keeping the party going. Without it, cells become disconnected, isolated.

A breakdown in the communication network can occur at any step. An excess of growth factors due to the inappropriate activation of a proto-oncogene bombards the cell with signals for growth, to which it responds with countless rounds of cell division. Or, the receptor itself might be defective, emitting a stimulatory signal for growth even in the absence of a growth factor. The outcome is similar. The cell is deluded by the misfiring of the receptor, and reacts according to the pre-determined rules by moving from one cell division into the next without a pause.

The proteins of the *ras* family of proto-oncogenes attracted much attention, and considerable effort has been devoted to finding out how they cause a malignant change. The *ras* protein belongs to a complex of G proteins responsible for transmitting biological signals from the cell surface to the chemical messengers on the inside. Normally, when a receptor receives a signal, it transmits it across the cell membrane, and once the message is sent, the system shuts down and returns to a "ground" state. When another signal reaches the cell surface, the system is again fired and the entire process is repeated.

A mutation in the *ras* protein causes the system to be trapped in an activated state, with a continuous, unregulated transmission of signals. This constant barrage overrides the normal controls, and the cycle of cell growth goes on and on.

Deep in the cell is the nuclear clearing house of all the information received by the cell. Depending on the connection made, the cell remains quiescent, continuing to work away on its schedule until further instructions are received. Or, it may enlarge, replicating its DNA at a great pace, and then divide. For this kind of scheme to work well, the cells are required to obey, to the letter, the rules. Any ambiguity in the instructions or straying away from the precise programme, agreed to from birth, will introduce such noise that the conversing among cells becomes unintelligible. This is dangerous for the cell.

The proteins of the proto-oncogenes do not function in isolation, but there is a constant interplay among them. The formation of a full-blown cancer occurs in several steps, in which the normal cell is converted to a malignant one through a series of progressive changes. The activation of any single proto-oncogene, through a mutation or gene rearrangement, contributes to the eventual development of a cancer.

In 1969, Henry Harris began a comprehensive approach to the genetic analysis of cancer cells. He fused normal cells with cancer cells. The phenomenon of cell fusion allows the intermingling of the nuclei of different cells, the cytoplasm flowing easily from one to the other. The fused cell behaves as a single cell, but with two alien sets of chromosomes. Somewhat unexpectedly, the hybrid cell resulting from the fusion of a normal and a malignant cell is no longer malignant. This implied that the normal cell contains one or more genes that interfere with the full expression of malignancy, and led eventually to the discovery of a separate category of regulatory genes — the tumour suppressor genes. Loss of these critical genes in a cell can be of grievous consequence.

The proto-oncogenes act in a dominant fashion. When they are activated, they exert a direct influence on the function of the cell. In contrast, the tumour suppressor genes are recessive; both genes from the maternal and paternal chromosomes must be lost or crippled for them to be a threat. For this reason, finding them and understanding how they work have lagged behind.

The disease that has provided the most useful information about tumour suppressor genes is retinoblastoma, a rare eye cancer which usually develops in children by age five. Provided that the cancer is diagnosed early, it can usually be successfully treated, and many children are now cured. There are two forms of the disease; one is inherited, while the other is not. Those with the inherited form of retinoblastoma pass on the trait to their children. The children who inherit the disease are susceptible to several cancers in both eyes, whereas in those without a family history of retinoblastoma, only a single tumour in one eye develops, usually also at an older age.

The inheritance of the *rb* gene, as it is dubbed, is not, however, sufficient by itself to transform a normal retinal cell into a malignant cell. Persons with inherited retinoblastoma develop only a few focal tumours in a background of several million retinal cells. Even though all the retinal cells contain the *rb* gene, most continue to function normally; only a tiny fraction actually progress to a frank malignancy. This means that for retinoblastoma to develop, it is not enough for the retinal cells to harbour the *rb* gene, but some additional genetic event must take place.

In 1971, Alfred Knudson put forward the idea that retinoblastoma is caused by two mutations. He thought that the inheritance of the *rb* gene might be the first mutation, and during life, a second mutation occurred that transformed the retinal cell into a cancer. The two mutations that are necessary before a retinoblastoma can develop are, in fact, the loss or deletion of the *rb* gene on the two complementary chromosomes in the cell. Unlike the proto-oncogenes that must be activated, the two *rb* genes in the cell must be silenced, suppressed.

In 1986, Robert Weinberg at the Massachusetts Institute of Technology provided unambiguous evidence of the loss of the *rb* genes in retinoblastoma. At first, the *rb* gene was thought to be no more than a curiosity, as it was involved in an obscure cancer, but problems with the gene have now been linked to some common cancers like lung, breast, and bladder cancers.

Even more compelling evidence for tumour suppressor genes has come to light in the past few years. Friend and his colleagues at the Massachusetts General Hospital reported that they have identified the cause of a rare but devastating pattern of familial cancers known as Li-Fraumeni syndrome. Persons with this inherited illness are prone to six specific types of cancers, as well as showing some increased susceptibility to several others. The cancer cells have a defective gene called the *p53* gene. Soon after, a variety of mutations and deletions in the *p53* gene have been described in lung cancer, bowel cancer, and others, confirming the importance of the tumour suppressor genes in human malignancy.

The genetic jigsaw puzzle is being slowly pieced together. A small number of key genes control growth and development. Some, the cellular proto-oncogenes, code for proteins that promote cell division; others, the tumour suppressor genes turn the process off. Under normal conditions, both classes of genes, working together, enable the body to perform the critical function of replacing dead or defective cells. The mechanism is itself quite a good one when used with precision, admirably designed to provide the checks and balances for orderly growth. Any wandering from this will introduce grave hazards for the cells. It is not a system that allows for any deviation.

The proto-oncogenes and tumour suppressor genes are highly conserved throughout evolution. They are found in distantly related species from the plant and animal kingdoms. This was persuasively demonstrated by Wigler, who found a structural and functional equivalence of the *ras* proto-oncogene in human cells and yeasts, two species whose ancestors diverged and followed separate evolutionary paths over a billion and a half years ago. Without the *ras* proto-oncogene, the yeast cell will not grow. Yet, when the yeast gene is replaced with the human *ras* proto-oncogene, the yeast cells are able to grow. This is possible because the yeast *ras* proto-oncogene codes for a protein that is very similar to the human *ras* protein, which is, therefore, capable of filling in for the absent yeast *ras* protein. The conservation of the regulatory genes is one of the remarkable discoveries of modern biology.

It is most unlikely that the similarity of the *ras* proteins in these two species could be a chance occurrence. Crick calculated the likelihood that amino acids, and by extrapolation the bases in DNA, will be arranged in a precise way to code for a single protein. Consider a protein molecule made up of 200 amino acids, a relatively small protein. Since there are twenty different amino acids, then at each site in the protein molecule, there are twenty possibilities. The number of possible arrangements of all 200 amino acids in the protein is 20 multiplied by itself 200 times; that is, 20^{200}. This is approximately equal to 10^{260}; written in long hand, this is the number one followed by 260 zeroes!

A far more likely explanation of the similarity of the human and yeast *ras* proteins is that their genes descended from an ancestral gene, which is modified slightly over evolutionary time span. Such divergence from a common progenitor gene, first proposed by Doolittle, is a consistent theme in nature. It is easier for species to duplicate, and then modify genetic components to suit their requirements than it is to assemble new genes *de novo* from random beginnings. It works. The uniformity of living things can be traced to their origin from a single ancestral cell. It is from this parent we take our looks; the resemblance of the proteins of the yeast to those of a human is a family resemblance.

The proto-oncogenes and tumour suppressor genes operate in tandem. The development of a full-blown cancer occurs in several steps with damage to both classes of genes. Mutation in a proto-oncogene accelerates cell growth, whilst mutation in a tumour suppressor gene releases the brakes that normally regulate cell proliferation. For many of the common human cancers, including those of colon, breast, and lung, the early event is a mutation of the *p53* gene, which causes the cell to ignore the various messages that are received from the outside and transmitted to the nucleus. Subsequent changes in other regulatory genes spur cell growth and a burgeoning mass accumulates.

As further genetic mutations occur, the cell becomes more and more virulent, disobeying the usual laws of territoriality. It is eventually emancipated from normal regulatory mechanism, and is on its way to becoming a cancer. A vortex of events is set in motion. The cell plunges into frank malignancy. Billions of cells are spawned, invasion of surrounding tissues takes place, and some break away and set forth in the bloodstream.

CHAPTER 5

SURVIVAL OF THE FITTEST

A remarkable phenomenon in Nature is that all cells increase their number through the process of mitosis. Single cells divide into two, then four, and so on. The original cells vanish totally into their progeny, but their genetic material lives on eternally, carried forth in their descendants. This arrangement works well enough for the simple one-celled organisms, like bacteria or amoebae. As long as there are sufficient nutrients in their environment, the constant renewal and replacement keep on going.

The fundamental mechanism of cell division may be the same for an amoeba as it is for a human, but here the situation is incredibly more complex. We can readily begin to grasp the scale of this when we consider that the adult person is made of a mind-boggling number of cells, about fifty trillion. More than just sheer numbers, the cells behave as a co-operative society, communicating with each other, working collectively instead of being detached, selfish. The entire enterprise would be practically ungovernable were it not for the regulatory genes. There is nothing to match the elegant organization of cells into a working, functional whole.

The trillions of cells are ceaselessly putting bits of information together, making connections, and responding. All of this is almost routine, workaday details. The fascinating aspect of this complex behaviour is the way it is done. The cells cannot be given prearranged instructions to be followed on a rigid, inflexible schedule. It does not work this way. Instead, the process is a dynamic one, responsive to changes, adaptable. If the liver is injured, the damaged cells are removed and replaced by new healthy ones, as those nearby divide to put everything right again. When this is done, cell division stops and the cells resume their everyday business.

We can think of the life of a cell in terms of a cycle. A cycle denotes a repetitive series of events with the final event terminating at the beginning of the first event of the next cycle. An everyday example of a cyclic event is the way the four seasons follow each other, coming full circle in a year.

As a cell goes through a cycle, it passes through several stages, dividing into two new daughter cells. A convenient starting point in following a cell through its cycle is mitosis, in which the cell duplicates its genetic material and distributes an equal share to each daughter cell. After a definite interval, each daughter cell passes through another mitosis, and the cycle continues.

Cells that continuously move from one cell division into the next are cycling cells. They produce more of themselves without a pause. The majority of human tissues contain cycling cells, which replace those that are steadily worn out and die. In the bone marrow, new blood cells are being constantly produced to replenish old ones, which live out their life in circulation before disintegrating.

But not all cells are in continuous cycle. Some stop dividing and become quiescent. They can, however, be recruited back into the cell cycle. These are resting or non-cycling cells. Cells of the liver are usually in a resting phase. They go about the day's business without much evidence of preparing to divide. Instead, their energy is channelled into protein synthesis. But if the liver is damaged, the remaining healthy cells react, re-entering the cell cycle to make up the loss. As soon as the injury is repaired, cell division comes to a halt, and the cell subsides once more into dormancy.

Other cells may undergo a number of cell divisions, but then leave the cell cycle permanently. These cells are unable to duplicate their genetic material and produce new cells. They are destined to die without any further cell division. The nerve cells are a good example of this type of cell. In post-natal life, they do not usually increase their numbers.

The life cycle of a cancer cell is quite similar to that of a normal cell. Even when the cell cycle of the mammalian cell was first described, the cancer cell was regarded as intrinsically different. Its growth was unrestrained, and it seemed to have it in for us in the worst way. On closer examination, however, this turns out not to be the case. Not only does the malignant cell go through the same phases of the cell cycle as the normal cell, but it also takes a longer time to do so. In the midst of what seems a complete breakdown in order, the same biological process is at work.

If the cell cycle time of a cancer cell is, in fact, longer than that of a normal cell, the old notion of how a cancer grows was called into question. Rapid cell growth was obviously not the whole answer. The predominant abnormal feature of the cancer cell is that it is constantly moving from one cycle to the next, not heeding the usual stop signals in place. This arrangement is clearly not mutually profitable. The overstepping of the line by the cancer cell inevitably leads to an infringement of biological borders, the pushing aside and destruction of normal tissues.

In the mid-1970s, biologists realized that the vast majority of cells in the body have a limited life span. They are programmed to undergo a finite number of cell divisions, clocked by their genome, and then they die. We are being continually renewed. Old cells lose their zest, and as they go down, one by one, their place is taken by new ones. It is a constant parade. If this is the way it is, then some mechanism must be in place to replenish the stock as it withers.

The whole process is centrally run. We start our lives as a single cell, derived from the coupling of a sperm and an egg, and before long billions of cells are spawned as the embryo becomes organized. At a certain stage, there emerge single cells which will have as all their progeny the various organs of the body. With exquisite precision and timing, they distribute themselves, differentiate, and work in unison with those of their own kind. They form the different tissues. At the heart of each tissue, a small proportion of cells is retained, separate, aloof from the differentiated ones. They have the potential to divide indefinitely and restore the stock. They are the stem cells.

The descendants of the stem cells become highly specialized to perform specific tasks. The thyroid gland cells produce thyroid hormones, the red blood cells carry oxygen to the body's tissues, and so on. After a limited life span, they die and, as they do, they are replaced. The source of the new cells is the stem cells.

The vast production of cells from a relatively small pool of stem cells is possible only because of a unique property. The rules are different for the stem cells. Unlike their descendants, they work on their own timetable, dividing when necessary, but not relinquishing the capacity for infinite cell divisions.

There is more to it than simply the potential for countless cell divisions. The stem cell pool is created from the omnipotent fertilized egg very early in embryonic development, and in the adult no new stem cells are generated from this original source. The pool is maintained by the special ability of the stem cell to replace itself. In the absence of this self-renewal, the pool could be quickly depleted in the continual process of new cell production.

Throughout life, a small stem cell pool is preserved. When a stem cell divides, it gives birth to two daughter cells, each following a different route: one retains the potential for self-renewal, stubbornly sticking to one purpose. In contrast, the other enters a *cul-de-sac*, differentiating into a mature cell of limited life span.

The system works well. For the constant and, in some tissues, surprisingly rapid turnover of cells, the integrity of the organism is entrusted to a small pool of stem cells. It allows central control. There are some fifty trillion cells in the body, and in the face of such overpowering numbers, Nature has delegated final authority to a small number of stem cells.

Since one cell cannot produce more than two daughter cells at a time, to achieve the immense number of cells in the adult from a single egg requires many cell divisions and many generations of cells. At the minimum, there must be at least fifty trillion cell divisions. But as cells are continually dying and must be replaced, the final total is much larger, conservatively estimated to be in the order of a thousand trillion.

The stem cells are responsible for the homoeostasis of the organism. One obvious way the system operates is through the amplification of the stem cell's progeny. This can be enormous. The daughter cell that differentiates into a mature cell may divide some 10 to 12 times before cell division ceases. The initial cell gives rise to two cells, both dividing further into 4, then 8, 16, 32, 64, 128, 256, and so on, in a geometric fashion. After ten rounds of cell division, about 1,000 cells are produced; after twelve, there are 4,000 cells.

It is estimated that the stem cells undergo only about 1,000 to 2,000 cell cycles during life to maintain the fifty trillion cells that make up an adult. It is an ingenious design. The stem cells, which are ultimately responsible for the well-being, the wholeness of the organism, are protected, electing to remain smaller and going through far fewer cell cycles than their progeny.

The regulation of the cell population is achieved in two ways: the control of the stem cell proliferation, and the control of the amplifying divisions that take place when the cell leaves the stem cell pool. It is the nature of living things to have such built-in safeguards. Of these, control of the stem cell population is the more critical. Once a stem cell becomes committed to differentiate, its fate is sealed. It goes through a pre-set number of cell cycles and then it dies. A mistake at this level can, therefore, be corrected in a short time. In contrast, the stem cells are "immortal", capable of an infinite number of cell divisions, so that an error here is not minor, not as easily redressed.

Under normal conditions, the stem cell proliferation and the expansion of its lineage are closely integrated and carefully co-ordinated to meet the needs of the organism. It is a tidy, stable scheme. There is a finely tuned balance between cell proliferation, differentiation, and cell death. As differentiated cells live, work, and die, new ones move in to take their place, the descendants of the stem cells. Every cell that comes alive is in trade for one that dies, cell for cell.

A great deal was learned about the incessant cycling of cells from some remarkable studies of the bone marrow cells. They are engaged, continually and automatically, in the production of new blood cells, which on reaching maturity take their leave and set forth in the circulating blood where they live out their lives, bound by their genetic instructions, compulsively carrying out their special tasks. The white blood cells roam around, seeking out invading micro-organisms which they surround, attack and bombard with an astonishing array of biological arsenals. The red blood cells are more complacent, simply concerned with the delivery of oxygen molecules from the lungs to all the body's tissues. The platelets collect at the site of injury, sealing off any haemorrhage. They all pass through their limited life in the single-minded pursuit of their delegated duties. Then they disintegrate, broken into tiny bits from which new cells are created in the bone marrow. It is a finely balanced enterprise. Cells die all of the time and new ones arise in the same volume to take their place.

There is a harmony in normal tissues between the loss of cells and their replenishment. But in malignant tissues, the stability is undermined. The progenitor cells of a cancer do not heed the regulatory drives that give the normal tissue its durable order. It is a distortion of the careful design put together over the long evolutionary span since single, separate cells got together for the construction of metazoans, and leading eventually to humans.

The stem cells of a cancer replicate at an excessive rate and their descendants show abortive attempts at differentiation. What results is a freakish caricature of the normal tissue. Clinicians have been aware for some time that different cancers display a range of traits, from unbridled growth to indolent chronicity. It now appears that this wide biological variability can be accounted for by the degree of cell proliferation and differentiation that is going on in the cancer.

Where proliferation is the dominant driving force with little or no evidence of differentiation, the system becomes unstable, and the progenitor cell pool expands relentlessly. Such a cancer is virulent, growing at a rapid rate and, unchecked, almost always fatal. Burkitt's lymphoma is a good example of this form of cancer. The tumour grows fast, almost doubling in size in a matter of a few days. The cells remain immature with little suggestion of differentiation.

In the more common situation, the malignant cells show some differentiation, albeit an inefficient and inadequate attempt. They lose some of the proliferative thrust, the degree varying with how far down the path to full differentiation the cells travel. In the extreme, we can imagine a case where all the malignant cells become differentiated. In so doing, they relinquish their self-renewal potential. Such a cancer is of little threat as the specialized cells are incapable of sustaining further growth. It falters and dies. In some unusual circumstances, a recognized cancer can regress, leaving little trace of its existence. This phenomenon of spontaneous regression is best described in superficial malignant melanoma of the skin. One possible explanation is that the signal for differentiation of the malignant cells occurs early in the life of the cancer. Unsustained by a stem cell, the cancer extinguishes quickly.

In 1976, Peter Nowell proposed a model of the evolution of a cancer that goes far in tying together many of these concepts. He recognized that a cancer arises from the accumulation of genetic changes, causing the malignant cell to proliferate uncontrollably. As more and more cells come into being, further mutations occur. Not all of these necessarily equip the cell for survival and, perhaps, many die. An occasional mutant cell may, however, acquire a selective advantage over the original parental cell, and assume the role of the progenitor cell whose progeny then forms the dominant population. This concept is shown in Figure 6. Over time, there is further selection of progressively more virulent lines. The whole process spins irretrievably out of control. New mutants invade and destroy surrounding healthy tissues. Others spread throughout the body. It is a despairing situation.

During normal growth and development, the descendants of the stem cells move through an orderly and deliberate path towards specialization, giving up the potential for self-renewal on the way. After a certain stage, there is no turning back of the differentiating cell, its course is set. It goes through several rounds of cell division, ages, and dies. And its place is taken by a new cell. Order is maintained.

The cancer cells are not in the same fix. The rules are lax, and order is lost in the frenzy. The progress of the cell along a restricted path towards specialization is blocked at different steps and alternative paths may be taken. Mutations accumulate upon mutations. And through it all, the cells hang on to their replicative power, grudgingly giving up any claim to their

Figure 6
Nowell's hypothesis of tumour growth

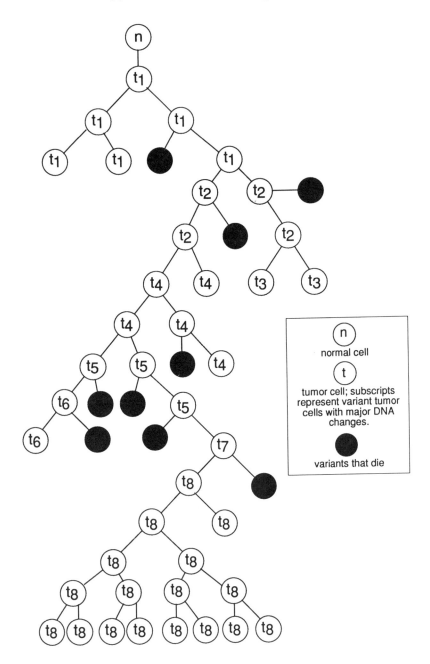

self-renewal potential. This seems to be the most formidable force at work. It is an almost ungovernable biological mix-up.

A cancer starts out as a mutated cell, and after a series of 30 cell divisions, there are about one billion cells, as each cell division doubles the number of cells: 1, 2, 4, 8, 16, 32, 64, 128 .... At this stage, the cancer is just one cubic centimetre in diameter, a comparatively small tumour by everyday clinical standards. What this means is that the opportunity for genetic changes in an instrinsically unstable system is astounding. Even if many of these random mutations do not propel the malignant cell, there is a good chance that the necessary few would occur.

When a normal stem cell divides, it passes on its genetic material to its daughter cells. As these differentiate, only the genes for their special trait are switched on; every other potential property remains latent, and the option of being anything else is denied. In a cancer, this tidy arrangement is abandoned, and various genes are switched on in disarray. An expanding mass of cells collects with aborted patterns of differentiation, incapable of serving any beneficial function, of carrying out any useful task.

Although the malignant cell does not make it down the path to full specialization, this does not appear to put it at a disadvantage, as it shows a propensity to cross over to alternate paths. Different genes are switched on in an erratic, disorderly fashion, causing the emergence of a diversity so distinctive of the cancer cell. A glimpse into this comes from the frequent finding that the cancer cell produces an array of proteins. The cell manifests a range of traits not expressed in its normal counterpart; in fact, some of these are found only in the embryo.

Lung cancer cells secrete several hormones. These molecules are not unique, but are produced by the thyroid, adrenal, or other glands. Liver cancer cells often produce a protein, alfa-fetoprotein, that is manufactured only during fetal development. The high levels of alfa-fetoprotein reach a peak between 12 and 15 weeks of gestation, but drop steadily after this. About 12 months after birth, the protein is barely detectable in the blood. The level can increase significantly, sometimes in excess of a thousandfold, in many cases of cancer of the liver.

The anomalous production of proteins is a special property of malignancy. It indicates a versatility. Genes that are active in fetal life, but are switched off after birth, can be recalled into service. Or, proteins elaborated by other types of cells are produced by the cancer cell. The co-ordination that allows things to go well is now open to tampering. Digestive enzymes break down the territorial barriers between cells, and allow a frontal assault by the malignant invaders; retractile proteins manoeuvre the cells through the bloodstream to other organs.

The versatility of the cancer cell is telling of the underlying instability that kindles the process. But the malignant cells make no molecules that are not found in some types of normal tissues, adult or fetal. This should not be unexpected, for they share the same genetic information with the normal cell. In its progression to malignancy, the cell does not acquire new genetic material. What is peculiar about the cancer cell is that the gene expression follows no pattern. While in the normal cell, most of the genes remain latent, they are switched on in a seemingly haphazard manner in cancer with no clear-cut recognizable goal.

If the cancer cell does not produce any new kind of molecules or possess unique metabolic mechanisms, we must look elsewhere to understand why these cells are so deadly. An obvious way to approach the problem is to figure out the traits that allow the malignant cell to cross the line. There is a continuous accumulation of cells into an ever-expanding cluster that infiltrates and sends forth emissaries, which eventually take hold and grow in distant organs. This aggressive behaviour is sustained by rampant cell division.

Unless a cancer cell divides, it poses no threat. But cell division is an indispensable attribute of normal cells. Nor does the rate at which the malignant cell divides provide a lead in solving the problem. In fact, during embryonic life, cell division occurs at a rate faster than in most human cancers. We are no further ahead.

The hallmark of a cancer is its ability to invade and spread. Again, this is not an abnormal biological behaviour. The developing embryo becomes implanted in the uterine wall of the mother, into which it infiltrates and establishes itself. Within the growing embryo, cells migrate, move, and roll into layers as the tissues are laid down.

Cancer cells are not cells that possess properties not available to some normal cells. No malice is intended. What appears to distinguish them is their unique ability to call at will upon the vast genetic resources, an ability that gives them an unfair advantage over normal cells, which are out-distanced in the race for survival and reproduction. It is every cell for itself, and the malignant cell, with no respect for the laws of organized growth, overpowers the normal. Without some strict order and regulation, the high cellular organization of humans would not be possible. When this order breaks down, a cancer is born.

Cancer kills. It kills by starving the host, by consuming too much of the available nutrients, and the situation only worsens as the cancer grows. The host is weakened, and finally succumbs. It kills by damaging vital organs, such as the liver, lungs, or brain, whose cells are pushed aside and replaced. And it kills by releasing toxic substances or causing a derangement of normal metabolism with which the body cannot cope.

The cancer cell has the versatility to flourish under conditions inimical to the normal cell. In a disturbing way, it is the fitter of the two. Able to call upon the immense genetic memory of its evolutionary experience, so faithfully encoded and stored in the DNA, it can switch from one biochemical pathway to another to suit its needs.

CHAPTER 6

CONTRARY TO NATURE

The most remarkable achievement of Nature is the invention of DNA. Around four billion years ago, in a bolt of lightning, as the primitive earth cooled, the earliest ancestor of DNA emerged in the ponds and oceans, a humble beginning of today's sophisticated forms of life. The first molecules made crude copies of themselves, competing for building blocks. As time passed, some became more and more efficient at this and joined with others to form a kind of molecular collective, the first living cell. After a billion tumultuous years, one-celled plants came together to form a colony, the beginning of multi-celled organisms. What followed was an equally rich and splendid history of the evolution of life. And strung through every cell are the descendants of that original master molecule of life. We are, all living things, close cousins.

DNA is highly conserved. As living organisms became more advanced with the organization of cells into complex mammals, the ancestral genes are not lost. New genes arise as extension and elaboration of the old ones. But once a species has adapted to a stable environment, major rapid changes do not take place. The capacity to stay on track is the real marvel of DNA. Random, spontaneous accidents or mutations do occur, however; it is what drives evolution. Without this, we may not have progressed far beyond the primitive prokaryotes, like bacteria or blue-green algae. Viewed individually, one by one, each of the mutations, each improvised new gene that slowly brought us along represents but a small embellishment. Wide swings are not tolerated if a species is to survive and not just flicker out.

A paradox does arises. We are, each of us, made of a fifty trillion cells, possible through at least as many mitoses, as they all eventually come from a single fertilized egg. When we take into account the large number of cells lost during life, the number of mitoses is even larger, probably in excess of a thousand trillion. Copying errors are more likely to occur during the duplication of the DNA as the cell goes through mitosis. Mistakes are uncommon, occurring once in every hundred thousand to a million mitoses. Since every gene is present in pairs, one maternal and the other paternal, the likelihood of a mutation in both pairs is this number multiplied by itself; that is, once in ten billion to one trillion mitoses.

Considered in light of the sheer frequency of mitoses in an average lifespan, even this seemingly low rate of double mutations is disquieting. We would expect such an event to occur between 1,000 to 100,000 times in a lifetime: this works out to be about 20 times a year to five times a day. If this is true, then making genetic mistakes is almost a routine part of living. We are programmed, coded, for errors. Yet, it cannot be so. There must be an insurance against such blundering.

Under normal conditions, the vast production of cells is, indeed, carried out in such a way that enables the organism to avoid accumulating mutations. In tissues where the cells turn over constantly, the new cells are the descendants of a stem cell, which will grow and differentiate into a large progeny, often reaching up to 1,000 cells after just ten cycles of division. In this process, the stem cell divides in such a way that one of its immediate daughter cells takes its place in the stem cell pool, whilst the other goes forth and spawns the expansive family. There is a built-in protection here. For a mutation to produce the wracking and unhinging of this ingenious design, it must be retained in the stem cell. Should it be introduced into a differentiating cell, programmed for eventual extinction, the mutation is lost when the cell is discarded.

The organization of cells into a hierarchy is an elegant and workable device by Nature to exercise control over the constant stream and jostling of cells. The stem cells, small in number, remain secluded, privileged, and protected. They ensure the long-term continuance of the organism. Their descendants, the differentiating cells, go about the day's business without much evidence of thought for tomorrow. Their numbers can be regulated to suit the immediate needs of the organism, and after a while, when they age and wear out, they are replaced, on schedule, by younger cells. It is a splendid and effective short-term strategy.

The rules are broken in cancer. The underlying order is lost and the regimented sequence of cells in the differentiating path is in disarray. The regulatory genes are switched on and remain stuck in this position. The forces that govern growth and development collapse all at once. In this raddled state, the more pernicious cells take over as the progenitor. Mutations, some of which provoke rampant cell proliferation, are preferentially retained.

There is a fundamental stability of DNA. It resists change. Mutations are most likely to occur during mitosis when the two strands of the molecule separate, each serving as a template for the synthesis of another strand. This results in two molecules of DNA, each of which is made up of an old strand from the original molecule, while the other is a new strand. If a copying error does occur, it is most likely to be in the new strand. As the mutation is present initially only on one strand, on the next round of DNA replication, the normal strand acts as a template to produce a fully normal DNA molecule, whereas the "mutant" strand gives rise to a new DNA molecule that contains the error in both of its strands.

The stem cell might be shielded against mutations due to copying errors of the DNA. Cairns postulates a mechanism that ensures the "immortal" stem cell always receives the older of the two parental strands, while the "mortal" differentiating cell receives the younger strand, the one in which the mutation is more likely present. His idea is depicted in Figure 7. By passing on any mutation due to copying errors in the DNA strand to the daughter cells that have a limited life span, the mutation is lost when they die. The stem cells, however, remain genetically pure.

The disorganized pattern of cell proliferation in cancer undermines this built-in stability. The malignant cell is exposed, it takes on new mutations, and passes them on to each new generation.

In the biological game between the normal and malignant cells, the rules seem much more flexible for the aggressor. It is more versatile, able to generate new variants with alarming facility. The normal cell is out-manoeuvred with only the winner staying at the table. A disheartening thought.

Figure 7
Cairns' hypothesis

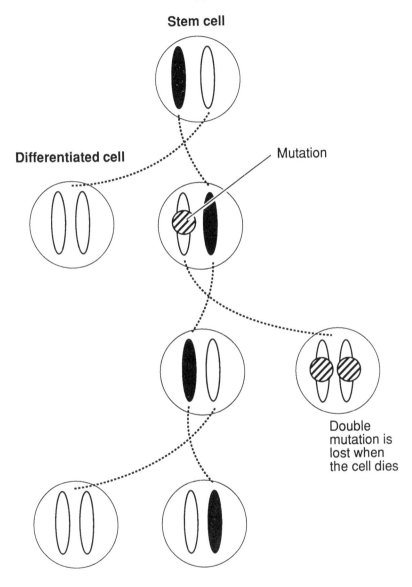

Parental strands segregate alternately to the right and left of the plane
of division, so that the stem cell keeps the same parental DNA strands
through successive generations.

Doctors now have a better insight into the intimate functioning of the normal and malignant cells. The mechanisms by which genes are switched on in the progression to cancer are understood in more detail, and with what has been learned in the past decade, the control of cancer appears to be within our grasp. The problems, to be sure, are still a challenge, but they are now approachable and can be worked on.

There is a public debate of the gains made in search for a cure, and beneath this can be sensed a pessimism and discouragement. Perhaps, it was premature to have expected much more until the underlying mechanisms of the process were fully understood. Until this is accomplished, designing effective measures can only be empiric.

We often lose sight of the fact that about half of all the major human cancers are now cured outright. By and large, this has been achieved by the skills of surgeons, radiation therapists, and chemotherapists. But advances are painstakingly slow. The optimism that greeted the introduction of cancer chemotherapy in the 1940s has now been tempered by the sad realization that most of the common cancers are not budged by today's drug therapy. Here and there, some successful inroads have been made, like the designing of multi-drug chemotherapy for malignant lymphomas, but these opportunities do not come often.

In trying to understand why this stalemate has been reached in the treatment of cancer, we come face to face with the intrinsic resilience of the malignant cell, not shared by its normal counterpart. The malignant cell can withstand the brunt of a chemotherapeutic assault that, at the same time, leaves the normal cell stunned. In fact, it is not unusual for cancer cells to display increasing, and often enormous, degrees of resistance to drug therapy, to which the normal cells, especially those of the bone marrow, become less and less tolerant.

This is a particularly difficult problem in cancer chemotherapy. Most anticancer drugs cause a transient lowering of the blood counts, which if severe can result in harmful infections or bleeding. Because of this, the drug dose is kept at a level to avoid too deep a fall in the blood counts, which over time do recover. But with repeated cycles of chemotherapy, there is a tendency for the blood counts to fall even lower and recover over a more protracted period. At the same time, cancer cells devise ways to circumvent the effects of the drugs, and continue their relentless growth. Nothing can be more frustrating in clinical practice than to observe the regrowth of a cancer after an initial favourable, and sometimes dramatic, improvement with drug therapy.

An explanation of the relative susceptibility of some normal tissues to chemotherapy is that their rapid proliferating cells provide a target for the anticancer drugs. This may seem an intuitively plausible explanation, since the drugs interfere with DNA synthesis and cell division. Clinical observation also appears consonant with this, as the actively dividing cells of the bone marrow and the lining of the gut often suffer the most damage from chemotherapy. A closer examination, however, discloses some difficulties with this explanation. For example, the use of a prolonged intravenous infusion of 5-fluorouracil, a common chemotherapeutic drug, has little effect on the bone marrow, but can cause severe ulceration of the digestive tract. Conversely, the cells of the digestive tract are relatively unaffected by cyclophosphamide, another common drug, which can cause life-threatening bone marrow toxicity at high doses.

Chemotherapeutic drugs exert different side effects in different tissues. This cannot be due solely to the proliferative state of the cells, but other mechanisms must be at work. Yet, whatever the cause for such differences, the normal cells exhibit a constant, and even increasing, susceptibility to their adverse effects. This contrasts sharply with the malignant cells which become progressively less susceptible.

An issue of considerable practical importance is the origin of this resistance to a wide range of anticancer drugs in malignant cells. This is brought about by cellular proteins. What this means is that the DNA contains the base sequences that code for proteins which enable the malignant cell to avoid the threat. Such genes do not arise *de novo*, but are present, conserved, in the cell's genome where they remain silent. The cancer cell, when confronted by poisons, turns on every defence at its disposal. It escapes. But the normal cell, although containing the same genes, appears unable to reach into and tap these remote, and possibly more primitive, regions of the DNA. It succumbs.

The presence of genes that code for the proteins responsible for drug resistance in cells implies some past evolutionary significance, and it is interesting to speculate what this might have been. The development of cancer chemotherapy has been based, in a sense, on trial and error, and occasionally sheer luck. It is from this sort of empiricism that we find the clue for these genes. Virtually every known plant extract, many microbial products, and solutions of some metals, have been screened at one time or another for anticancer properties. From this, an inventory of 40 to 50 drugs has been identified for clinical use. The common denominator in all of these is their ubiquitous presence in the environment. The anticancer drugs resemble naturally occurring substances.

James Goldie at the British Columbia Cancer Agency speculates that over the evolutionary time span, the progenitors of today's animal species have adapted to these environmental substances, an adaptation that might have given them an edge in the game of reproduction and survival. The genes of the ancestral organisms are not lost, but are retained and passed on indefinitely. For the most part, they remain latent, buried in the deep recesses of the cell's nucleus. It is quite possible that the variants that emerge in the malignant state represent the resurfacing of the repressed segments of DNA. The very mechanisms that ensured the success of our ancestors in the distant past, and are so faithfully conserved in our genetic library, may now present a formidable barrier to the eradication of the scourge of cancer today.

In 1989, Victor Ling at the Ontario Cancer Institute in Canada, reported the first direct and compelling evidence for the presence of proteins in malignant cells that permit them to evade anticancer drugs. Ling described a single protein molecule, the P-glycoprotein, in the cell membrane that pumps drugs out of the cell. By preventing their accumulation, the P-glycoprotein protects the cell.

It now turns out that the P-glycoprotein is found in the normal cells lining the gut. It could be a protective mechanism. The diets of animals contain a large variety of organic and inorganic substances of diverse chemical structure, which can be harmful if they get into the body. The P-glycoprotein serves as a first line of defence by pumping them out. The scheme is itself a good one, when used with purpose, and admirably designed for dealing with the flood of chemical compounds through the alimentary canal.

Most types of human cancers have a surprisingly large amount of the P-glycoprotein in their membranes, primed to fire the moment a toxic molecule strays into range. There is something like an aggression in the activity, but it is a form of self-preservation. Here lies the obstacle to successful cancer chemotherapy. The malignant cell unlocks the vast genetic reservoir, bringing forth in what seems a collective derangement, random traits, some normally expressed only under strict control, others, perhaps, repressed for millions of years.

A sperm and egg can fuse and eventually grow into a person; the cells multiply and differentiate, becoming fifty trillion strong. The amazing feat is that everything falls so neatly into place. There is nothing to touch the spectacle. All the information needed for this is contained in the chromosomes. In every cell, genes are switched on at just the right time; all others are switched off. The precision is awesome. It is the powerful legacy of four billion years of evolution.

But when things go wrong, the whole enterprise tumbles wildly out of control. Genetic bits of information are scattered about, torn to shreds, disintegrated. It is, as Galen remarked nearly two thousand years ago, "Contrary to Nature".

CHAPTER 7

THE FOURTH FRONTIER

The most conspicuous benefit of modern medicine is that we are living longer. We watch, with each passing year, the medical advances that put death off. A century ago, every family could expect to lose members throughout the early years of life from some dreadful disease. Now, thanks to modern medicine, this is no longer so in the industrialized nations. The chance of a child born today of reaching the age of 65 is 70 per cent, while it was only 25 percent for children born around the turn of the century.

We now enjoy a reasonably good chance of a healthier, longer life despite the persistence of some major diseases like heart disease and cancer, or the emergence of new ones like AIDS.

We take much for granted today without realizing how comparatively new all of this is. Until the 1830s, the greater part of medicine was hopelessly confused, based on guesswork. Almost all of the treatments were designed through crude empiricism, snatched out of thin air. Not surprisingly, most did not work, but what was more astonishing was that once tried, some remained in use for decades or centuries before being given up. It was a despairing state of affairs.

The first major triumph of medicine came with the containment of the epidemics that spread explosively through the population, striking down their victims with indiscriminate abandon. In 1829, a new and terrifying disease appeared in Europe. It caused watery diarrhoea and vomiting, leaving the patient weak and dehydrated. In a matter of days, death ensued.

Cholera swept through Paris and left 7,000 dead in just eighteen days. But it was Britain, the most industrialized country in the world at the time, where the pestilence hit hardest. In the first two years, it killed over 22,000 people. The degrading living conditions in the urban centres were the perfect breeding ground for the disease. As people flocked to the cities in search of work in the mills and factories, they crowded into dirty, unhygienic living quarters. Most were underfed and chronically ill.

By the time cholera struck in England in 1831, the rivers were polluted with untreated sewage or with effluent from the mills and factories. The only other source of drinking water was communal wells, but these were not much better, and were also contaminated with wastes from the sewers. Clean water was a luxury denied the huddled masses in the cities. Under such appalling conditions, cholera left its ravages. There was a complete lack of understanding of the disease itself, and medicine could offer little in the way of assistance against it.

The cholera epidemic precipitated widespread riots throughout Britain, prompting the government of the day to set up committees to study the extent of the problem. In 1842, Edwin Chadwick published a wide-ranging report on the sanitary conditions of the working population of the country. In this, he documented that bad sanitation, polluted water supplies, and filthy living conditions shortened life expectancy. At the time, the average life expectancy among the privileged class was forty-three years, but for labourers, was just twenty-two. Moreover, a staggeringly high number of children under five were dying of diseases.

Chadwick's work was extended by William Farr, a statistician, who had studied in Paris, where, some twenty years earlier, the science of statistics had already enhanced the quality of medicine through the rigorous evaluation of the effects of therapy. When Farr analyzed where cholera had struck most severely, he came up with some interesting results. He noted no relationship between the upper and working classes of society, nor any correlation with location or occupation. The only meaningful observation was that the incidence of cholera decreased in relation to the height above the Thames at which the victims lived. Farr was unable to explain this, but was convinced that it was relevant. He suggested that the stench from the river was responsible.

It was left to John Snow, a London physician, to make the connection. He suspected that cholera was transmitted through the handling of food by persons whose hands were dirtied by diarrhoea or vomit. In 1854, he was proved correct. In London, six hundred people died suddenly after drinking water from a common well. Snow found that the well water was contaminated by sewage from a nearby cesspit. When the pit was sealed off and the water filtered, the problem ceased.

The case for improved public sanitation was championed by John Simon, the first medical officer of health for London. He advocated public health measures, began routine building inspections, and instigated a series of reforms to ensure clean drinking water. By 1858, the London sewage system was renewed and cesspools abolished. All sewage was carried by pipes downstream from the city to a point where it flowed out to sea. With these measures, cholera vanished.

At about the same time, Louis Pasteur had demonstrated that germs caused fermentation of milk and wine. Until then, the reason that infection occurred in hospitals and communities was unknown. Following Pasteur's lead, Joseph Lister applied antiseptic techniques in the operating theatre, and deaths from infections contracted in hospitals dropped like a stone. Rapid discoveries followed that firmly established that micro-organisms cause diseases.

With the recognition of the role of bacteria in illness by the late nineteenth century, basic research began to work out the details of this connection. The major pathogenic organisms at the time, mainly the tubercle bacillus and the syphilis spirochete, were known, and the disabling, and often fatal, diseases they caused were thoroughly studied. This understanding of the infectious diseases enabled a sensible search for antibiotics.

The antibiotics were introduced in the mid-1930s with the entry of sulphonamides and penicillin. This is often cited as the beginning of modern medicine, but it does not tell the whole story. The search for compounds to fight off the infectious agents began some fifty years earlier. It is claimed that the discovery of penicillin is a classic example of serendipity in scientific investigation. In 1928, Alexander Fleming noted that bacteria growing in the vicinity of a contaminating mould, *Penicillium notatum,* in the laboratory, were killed. Almost a decade later, Howard Florey extracted a substance from cultures of the mould, and demonstrated its usefulness in the treatment of bacterial infections in humans.

At face value, it does seem as though biologists had, indeed, stumbled upon antibiotics by blind accident. Such, however, is a limited view. The crucial step is that the significance of the observation, however tangential, was grasped. What is needed for progress to be made is to act on the information. The capacity to make connections, to see the possibilities, is the secret.

The identification of pathogens and the study of the diseases they caused led to the concept of antibiotics with which we can rid ourselves of them. The research can be long and painstaking. It took about fifty years to go from the concept to the discovery of penicillin. It may seem that it all came in a flash, in a spectacular moment of inspiration, when Fleming first noted the strange phenomenon in his laboratory. But the stage was set for insights like this; it is the fitting together of things that no one had guessed at fitting. These are the good ideas in science.

The entry of antibiotics into the pharmacopoeia was another major occurrence in medicine. Its impact lay in the new ability of doctors to cure a great number of patients whose illnesses had previously been untreatable. There was nothing like it before. Previous therapy for diseases had little beneficial effect and might have been, in fact, downright harmful. With antibiotics, however, cures became commonplace.

With proper sanitation, the scourge of epidemics was contained, and with antibiotics most of the serious bacterial illnesses have left us. The other seminal event in medicine was the discovery of anaesthesia. Until its use, surgery was a distressingly painful experience from which patients were lucky to survive.

This changed with the use of anaesthesia that enabled doctors to carry out increasingly more complex surgical procedures.

The discovery of anaesthesia can be traced back to 1842 when the anaesthetic quality of sulphuric ether was first recorded by Crawford Long, a surgeon, in Georgia, United States. He performed an operation to remove a tumour on the neck of a patient, John Venable, who had breathed it. Over the next few years, Long repeated this practice in five other patients and published the results of his experiments in 1849 in the Southern Medical and Surgical Journal. Surgery has never been the same since.

One major problem remained. Although surgical expertise improved and patients survived the operation, they fell prey to infections. Blood poisoning, erysipelas, and gangrene were a genuine threat that dampened the advances in surgery. It remained for Louis Pasteur to lead the way in the development of antiseptic techniques to complete the revolution started by anaesthesia.

The history of modern medicine that was ushered in sometime in the 1830s is marked by three milestones: sanitation, anaesthesia, and antibiotics. There is no doubting the tremendous impact on our health.

We can learn important lessons from this period of medicine. We accept without much thought the cures of many of the infectious diseases with antibiotics. It is hard to imagine a time when tuberculosis was a great, perennial killer. Yet, this was how it was just fifty years ago. It needs thinking about. The dramatic change did not happen by chance. It took the concerted efforts of doctors and scientists over many years to create the store of knowledge and information that led eventually to the discovery of the drugs to kill the tubercle bacillus.

We remember little of the many years of hard work that went into the development of antibiotics. We celebrate, instead, its culmination, rejoicing in a new world of miraculous cures. We became therapeutic enthusiasts, and in doing so, neglected the prerequisite for progress. To turn a disease around successfully, we must first understand the underlying mechanism.

A complacent sense of optimism has dominated most of our thinking about cancer for a long time. In the early days of cancer chemotherapy, there was loose and careless talk of a "magic bullet" for cancer, and the idea still crops up now and again in medical literature and the lay press. It was a simplistic hope. Somewhere out there is a compound that will kill all cancer cells. All it takes is looking and, with the passage of time and a lot of good luck, we will find it. The unprecedented amassing of knowledge in biomedical science in the past several years has wrenched us away from this naive notion. A new view is slowly taking hold. We are rediscovering a concept that served us so well in the earlier part of this century. To control the relentless march of cancer, we must understand the intimate functioning of normal and malignant cells.

The discovery of the regulatory genes — the proto-oncogenes and tumour suppressor genes — has brought us further along this road than ever before. The role they play in the regulation of the cell is being brought into sharp focus, and with all that has been learned in the past decade, the problem can now be approached in a direct, sensible manner.

Much of the progress in biology over the recent past comes from the clever devices to cultivate mammalian cells in the laboratory. They can be manipulated, tricked, into revealing the exquisite details of their many working parts. Whole new blocks of information are being brought in at an increasingly rapid pace. We have come to take much of this for granted. Yet, what has occurred has the force of a revolution. The unimaginable a few years back has already begun. The genes, tucked away in a forbidden city in the deep interior of the cell, are now no longer beyond reach.

In 1980, Martin Cline, a haematologist in the University of California at Los Angeles, attempted the first human experiment in gene therapy. He was trying to find a cure for a common genetic disorder of the blood. The disease was thalassaemia in which there is a lack of a protein in haemoglobin in red blood cells, a deficiency caused by a mutation or deletion of a single gene. By then, the gene for haemoglobin was cloned and available.

Cline chose to conduct his experiment in Israel where thalassaemia was more common than in the United States. He extracted from two thalassaemic patients bone marrow cells, in which are the precursors of the red blood cells that circulate in the blood. Cline then treated the bone marrow cells with the gene for normal haemoglobin, and injected the cells back into the patient's bloodstream. From here, the treated bone marrow cells would find their way back home to the bone marrow where they reside and multiply. His hope was that enough of the treated bone marrow would take up the new haemoglobin DNA, and thus make the normal protein.

The experiment did not work. It was a bold and daring attempt for which Cline came under severe criticism for acting too early. Removing bone marrow cells and giving them back to a patient was a proven technique. On this, he was on firm ground. What was novel about Cline's treatment was to insert a new gene into the cells. At that time, the methods for putting DNA into human cells were quite inefficient, and most likely only a small fraction of the cells would have taken up the gene. Equally important, it was essential that the stem cells be the ones that picked up the gene. Cline treated all the cells in the extracted marrow with the hope that there would be sufficient stems cells in the sample, and that enough of them had taken up the DNA to make a difference. As stem cells make up just a small proportion of the cells, this was the serious flaw in the design of his experiment.

In the years following Cline's unsuccessful attempt at gene therapy, significant progress has been made. A method to deliver genes into cells was found. The new technology of gene therapy relies on an ancient, and improbable, product of evolution: the retrovirus. The viruses, packets of nucleic acids, already rent and share the cells of vertebrate animals, where they usurp the cellular machinery for their own purposes, replicating in their own fashion with their own genetic material. They tuck themselves up inside, replicate, and burst out again. Eventually, the immune system is alerted to the invasion and the arsenals for fighting off germs are mobilized. In the vast majority of cases, the virus is fended off; full recovery is the rule.

However, in uncommon instances, after an incomplete biological negotiation, some of the retroviral family occupy the cell, and establish themselves in the nucleus by edging their nucleic acids into the chromosomes. They remain here, protected, and privately follow their own agenda. They are, in a sense, mobile genes; they flit from one organism to another, carrying strings of nucleic acids. At times, they cause grievous harm. Some of the retroviruses are responsible for rare types of human cancers.

The trick is to exploit the retrovirus and use it as a vehicle to transfer genes of our choice. When a retrovirus gets into a cell, its genetic material takes up lodging in the chromosomes, but it takes along with it its own switches. The provirus uses the molecules that float around in the nucleus, but it maintains a private schedule, controlled and guided by its regulatory devices. Once the switch is flipped on, all the genes wired into its circuit are read.

The retroviruses have been studied extensively in the past several years. Biologists have identified those parts of the virus's genome that are involved in cancer, the parts that are required to make more viruses, and the parts that are responsible for its insertion into the cell's chromosome.

It would seem a hazardous business to use a potentially dangerous vehicle like a retrovirus to carry a piece of DNA into a cell. The strategy is to remove the genes that make the virus grow or are connected with cancer. In their place is put place a human gene. It is then no longer a retrovirus, but a hollow protein sphere in which are housed a new human gene and segments of the viral genome responsible for its integration into the cellular chromosomes and for regulation of its gene. When the greatly altered retroviral elements infect a person, the genetic material is released in cells where it links up with the chromosomal material. Under the supervision of the retroviral switches, the new gene is read, its message translated and carried forth throughout the cell. Any gene can be added, fitting neatly into a niche in the chromosomes.

By 1989, a protocol for conducting a gene therapy experiment was approved in the United States after extensive review by clinicians, scientists, ethicists, and lay persons. Permission was given to Steven Rosenberg and his colleagues at the U.S. National Institutes of Health in Bethesda, Maryland, to inject into patients, who had advanced malignant melanoma, blood lymphocytes into which a foreign gene had been inserted by a retroviral vehicle.

Some of the blood lymphocytes, the tumour infiltrating lymphocytes, respond powerfully to the presence of a cancer. When they sense a cancerous growth, the word is sent out. Other lymphocytes become activated and proliferate, expanding their numbers. They assemble at the site of the menace; they invade, infiltrate, and destroy the malignant tissue. It is a biologic struggle of the worst kind, with the winner taking all. Likely, this occurs all the time. The lymphocytes patrol our deepest tissues in search of the deviant cells of cancer, which are engaged in a showdown. When everything goes well, the malignant cells are outgunned. At other times, the outcome is unfavourable; the intruders gain the upper hand and repel the blockade.

The special ability of tumour infiltrating lymphocytes to attack and kill off cancer cells has been used to advantage. Human tumour infiltrating lymphocytes can be harvested from the blood of cancer patients and grown in the laboratory to increase their number enormously. When injected back into the patient, they circulate in the blood and home in on the cancer.

This technology is being used in the treatment of cancers, especially malignant melanoma. Tumour infiltrating lymphocytes cause significant reduction in the size of the metastatic cancer deposits, but only in a few cases is there complete disappearance of the cancer. In a sense, this is disappointing. Tumour infiltrating lymphocytes may be reliable in the surveillance and eradication of cancer cells as they arise, but are unable to cope with cancer that is already established. No amount of bolstering by sheer numbers can turn this around.

Nevertheless, the information gained from this research has moved medicine in a different direction.

A landmark achievement came about in 1990 when Rosenberg demonstrated that gene transfer can be accomplished in humans using a retroviral vehicle. The vehicle is derived from the Moloney leukaemic retrovirus, which causes lymphoma in animals, but is so greatly modified that it no longer contains any intact viral genes, and thus cannot replicate in the usual fashion. The gene selected for transfer coded for resistance to an antibiotic. This, therefore, was a probe into the efficiency and feasibility of gene transfer rather than true gene therapy. The gene becomes a permanent resident of the tumour infiltrating lymphocytes, and labels the cells and all their progeny for as long as they live. In this way, it becomes possible to track the tumour infiltrating lymphocytes, using its new gene as a marker to distinguish it from other lymphocytes.

This study provided the first evidence of the efficiency and safety of using a modified retrovirus to deliver a gene into a human cell. There were no ill effects of the gene transfer and no detection of harmful live viruses. The success of this experiment now offers doctors a platform to launch a new assault on cancer.

The next step is to send selected genes in tumour infiltrating lymphocytes to the cancer. The first gene chosen is that for tumour necrosis factor or TNF. TNF has a long history in biological therapy. It was observed that erysipelas, a streptococcal skin infection, could on rare occasions cause shrinkage of malignant tumours. In 1893, William Coley, a surgeon at the New York Cancer Hospital, inoculated ten patients with advanced cancers with erysipelas cultures, and observed that growth of the malignancy was inhibited. Some of the inoculated patients developed streptococcal infection, but all had fever even in the absence of a clinical infection. This suggested to him that an infection was not required for inhibition of growth of the cancer, and he subsequently used extracts of two bacterial species, *Streptococcus pyogenes* and *Serratia marcescens*, in the treatment of patients. Coley found that the mixture of bacterial extracts, called Coley's toxin, was as effective as inoculation of live bacteria. In either case, however, the treatment caused considerable side effects, and its benefits were transient and unpredictable. As a consequence, Coley's toxin lost its appeal as a cancer treatment.

About fifty years later, Murray Shear demonstrated that extracts of *Serratia marcescens* caused tumour necrosis when injected into mice with sarcomas. This work was expanded by Old at the Memorial Sloan-Kettering Cancer Center in New York. He discovered that the necrosis of the tumour was induced by a substance in the blood of the treated animals, which he described as tumour necrosis factor.

The TNF molecule is produced in response to infection by bacteria and infestations by some parasites. But its anticancer property is what attracted most attention. TNF has an interesting selective action against cancer cells but not normal cells. Its use in the clinic, however, was severely limited by a myriad of other toxic effects that complicate its use. A sufficiently adequate dose of TNF cannot be injected directly into the veins without evoking a violent, and often lethal, reaction. The body produces minute quantities of TNF to deal with intrusion of bacteria, but if it is overwhelmed by a large dose of TNF, it reads this as the worst of bad news. There is fever, the blood pressure falls, and shock ensues.

What was sought was a means to deliver TNF to the site of a

cancer, and so avoid the exaggerated reaction to an injection of high doses into the bloodstream. Because tumour infiltrating lymphocytes aggregate together in cancer, they are suitable for this task. Rosenberg and colleagues have successfully inserted the TNF gene into human tumour infiltrating lymphocytes which are then injected into patients. The approach is an ingenious one. It allows the delivery of molecular explosive devices in the midst of a cancer; it is mined.

The insertion of the TNF gene into human lymphocytes with a retroviral vehicle, and setting these forth to invade and destroy cancer is yet another of those astonishing feats of modern medicine. It is, no doubt about it, a major event, a triumph of biological science applied to medical practice. It might still be too early to assess what impact this particular experiment with TNF gene therapy has in the overall management of cancer patients, but a new approach has now been established. Doctors can select a gene, any gene, insert it into a human cell with the help of a retrovirus, and then release these engineered cells into the body.

Rosenberg is using the technique to deliver TNF to a cancer. This does not fix the underlying defect in the malignant cell, but offers the hope of combating the disease. A more direct approach to cure diseases by gene therapy is already under way. In 1990, Michael Blaese and his colleagues at the U.S. National Cancer Institute, started an investigation of gene therapy in children with an inborn enzyme deficiency. The disease is severe combined immunodeficiency or SCID. It is a rare genetic disease that is inherited as a Mendelian recessive trait; this means that the disease is expressed only if both the maternal and paternal chromosomes lack the gene.

SCID is caused by a lack of a single enzyme, adenosine

deaminase or ADA, which is important in the metabolic path-way of nucleic acids. ADA is present in most cells, but is partic-ularly active in the lymphocytes that circulate in the blood. In the absence of the enzyme, the normal biochemical process is thrown off track. The blood lymphocytes bear the burden of the genetic misfortune. Their number is lowered and their function as mediators of immunity severely compromised.

Up to 90 per cent of children with ADA deficiency run into trouble by one year of age. They suffer repeated infections and fail to thrive. Many also have bone deformities and sometimes neurological abnormalities. It is the recurrent infections due to their immunodeficient state that is the overwhelming concern of these children, almost all of whom die at an early age with-out treatment.

Until now, the treatment of choice has been bone marrow transplantation. By replacing the diseased cells in the bone marrow with the cells from a healthy donor, the enzyme levels in circulating blood cells are normal, and immunological func-tion is restored. While bone marrow transplantation is reason-ably successful in about 70 per cent of cases, unfortunately most children with ADA deficiency do not have a matched donor. In the absence of a suitable bone marrow transplant donor, other forms of therapy have been used. Some children have been injected with the ADA enzyme, but the success of this is limited.

The lack of a single enzyme in SCID means that its gene is missing. It is a freak of nature. The absence of a single gene, however, makes it an attractive model to explore a new form of therapy. Biology has moved from strength to strength, elaborat-ing new and better technologies, delving deeper and deeper into the cell. Genes are manipulated, cloned, spliced, and now in the ultimate test of the power of biomedical science, grafted into a living human cell.

By 1989, Dusty Miller and his group in Seattle, Washington, developed an efficient way to pack the human ADA gene into the Moloney leukaemic retrovirus. In the next step, lymphocytes are taken from the blood of affected children and grown in the laboratory to increase their number. There is nothing particularly novel about this; it is a routine, everyday ploy. Next, the cells are infected with the retroviral vector carrying the human ADA gene. The retrovirus, an unwitting agent in its new role as a therapeutic tool, enters the cell and delivers its genetic package. The human ADA gene, accompanied by its viral switches, elbows its way into the cellular chromosomes where it lingers on, looking just like another gene to the cell itself. The missing ADA gene in the lymphocyte is now replaced. As lymphocytes live for months, or even years, it is possible to correct the deficiency in the children for a long time, after which the treatment can be repeated.

In September, 1992, the research team at the National Institutes of Health held a party to mark the second anniversary of the ambitious experiment to cure SCID with gene therapy. Among the guests were two girls, aged six and 11, who had been injected with lymphocytes carrying the transplanted ADA gene. They were both born with the defective ADA gene, but now have a functioning immune system. Rather than being confined to a life in isolation to protect them from catching every germ around, they now attend public schools. It was an occasion worth celebrating.

The ways in which carcinogenic substances, or viruses, or a multitude of other factors still unrecognized, intervene in the working of a cell are problems that can now be approached in a rational manner. For too long, much of the thinking about cancer focused on outside causes for the things that go wrong. This led to the popular view that cancer is caused by environmental influences. Somehow, if we breathe clean air, eat wholesome foods, and shield ourselves from all the chemicals, man-made or natural, we would remain healthy and free from cancer. This is true to an extent. A major menace is cigarette smoke. Heavy smokers are more likely to die from lung cancer; there is no puzzle here; smoking is a dangerous thing to do. But other environmental factors, or exposure to occupational chemicals or pollutants, are much less straightforward. The plain fact of the matter is that for most common human cancers, with the exception of lung cancer, there is no definite predisposing outside influence, but several different factors seem to act in concert to derail the regulation of cell behaviour in the progression to malignancy. How these act and interact, one with another, is unclear, and it may well turn out that even when fully understood, they do not lend themselves to quick and easy solutions.

The thing to do is to make an all-out effort to unravel the underlying mechanisms by which the cell's genes are switched on or off as a cell turns malignant. Biologists no longer regard this as an impossible task, and are confident that with the tremendous surge of new information, they will soon find out how to control the errors in the process of cell growth and development. Just a few years ago, this would have seemed unbelievable; yet, here it is.

To be sure, gene therapy in human malignancy still faces some hurdles. The current methods of gene transfer with a retroviral vehicle result in the insertion of the incoming gene without correction of the existing mutant gene. The inborn genetic disorders, like ADA deficiency, are thus ideal candidates for gene therapy. In cancer, however, several genes malfunction. Proto-oncogenes misfire and tumour suppressor genes are silenced. At first glance, it seems that you cannot just step in and fix the problem. Different and diverse genes are switched on at the wrong moment, accelerating the proliferation of cells; and others that normally act as brakes are cut off. All of these, working together, drive the malignant process.

The realization that the malignant state is reached in several interdependent steps makes intervening with gene therapy more encouraging. Maybe not every gene that malfunctions has to be fixed. Instead, all that might be needed is to reach in carefully and correct just a few of the errant genes to foil the march towards malignancy.

There is the intriguing possibility of knocking out the expression of oncogenes with "antisense" molecules. Normally, only one of the two DNA strands in any given portion of the double helix is transcribed into RNA, and it is always the same strand. This is dubbed the "sense" strand; the other is "antisense".

A synthetic gene can be designed so that only the antisense strand is transcribed. The result is antisense RNA with a sequence of bases complementary to the normal RNA. Antisense RNA can bind to the sense RNA and prevent the cell from translating its genetic information into proteins.

Antisense compounds have already been crafted and doctors have begun the first clinical tests of their value to fight leukaemia. It is another example of how a deep understanding of the way a cell works is paying off. The antisense RNA strategy can be used to subdue the expression of wayward genes in the cell. It is a clever exploitation of the molecular blueprint that

Nature has drawn up billions of years ago, another triumph for biological science applied to medicine.

A more direct, frontal approach to cancer with gene therapy can now be contemplated. The issue here is: Can the aberrant genes of cancer be replaced by normal functioning genes? The appeal of this is that it tackles the problem at its root. It is the ultimate reward for the indefatigable efforts of the past two decades, since the cell has become a laboratory instrument that can be manipulated and studied in the same manner as bacterial cells were in the early 1950s. With considerable insight into the intimate functioning of the normal and malignant cell, biologists can determine the mechanism of normal growth and the ways proto-oncogenes and tumour suppressor genes misfire.

It seems that the more likely candidate for gene therapy is the family of tumour suppressor genes. Members of this family are responsible, in part, for the progression to malignancy when they are absent or crippled. The lost genes can theoretically be replaced by a new normal gene carried into the cancer cell by a suitable vehicle, such as a retrovirus. Once in place in the chromosome, it expresses itself, and passes on its copy to each new generation of cells. There is no reason why a second or third tumour suppressor gene cannot be inserted in tandem or separately.

The oncogenes might be a different problem. These genes remain active in the cancer cell. Those that are mutant could conceivably be replaced by a normal proto-oncogene, but it is uncertain whether this would be sufficient to reverse the process without first switching off or downgrading the mutants.

It is, perhaps, fortuitous that a full-blown cancer requires the activation of several proto-oncogenes and tumour suppressor genes. A single oncogene or tumour suppressor gene does not on its own inexorably lead to the formation of a cancer. It takes the collective action of several genes to permit a cell to make the transition from a state of normalcy to that of malignancy. The practical implication of this for the use of gene therapy in cancer is that correcting a few of the genes might be enough to stall the progression to fulminant malignancy.

At present, the most efficient delivery system for gene therapy is the retrovirus. It may well be, however, that other methods will become available in the future. The retroviral vehicle is what we have to work with now. This system is useful for inserting genes into cells that can then be injected back into the bloodstream. But a blanket insertion of genes into every cell in the body is undesirable; ideally, the gene should be put into the malignant cells only.

Many viruses have a fondness for certain cells. The cold virus is selectively attracted to the cells of the airways of the lung, while the hepatitis virus goes to the liver cells. There are many examples of this in nature, and we can learn from them. The selectivity of a virus for a cell depends on specific molecules or receptors of wildly variable configuration on the surface of the cell. Normally, they serve for the docking of molecules with a close fit. As molecules drift at random through the spaces between cells, they hunt for matching receptors, onto which they become locked. For a molecule to dock and couple with a receptor requires that the two have the proper interlocking surfaces. When this is achieved, they lock onto each other, pair up, and the molecule is swept into the cell. When the force of attraction is not strong enough for locking together, the molecule wanders off, sniffing for matching receptors, catching at everything until it finds a partner. This process of sorting and selecting allows each type of cell to latch onto specific molecules that are then admitted into its inner provinces; others are flung off.

A retroviral particle that contains a coat protein that recognizes only human white blood cells would permit the retroviral carrier to be given intravenously with little danger that cells other than the blood cells would be inadvertently infected. Such a retroviral vehicle could be used to treat diseases of the white blood cells. Similarly, specific viral vehicles for lung cells or liver cells can be used to deliver a gene into these specific cell types.

In the future, as more is learned on how to package DNA, an intravenous injection might be simpler and more direct. This approach was tried in France in 1983. DNA was wrapped in the fats and proteins that make up the cell membrane of red blood cells. Manipulating the outer coat could allow doctors to target a piece of DNA to a specific tissue. The French group was not successful in their attempt, but the technical difficulties might be solved.

Another problem to overcome is to direct the new piece of DNA to specific chromosomes. The optimal system not only would deliver the new gene into the cell of choice, but would also direct it to a predetermined chromosomal site. The incoming gene in the retroviral system brings its own switches, derived from the viral genome. If, by chance, it integrates itself next to a resident proto-oncogene, its independent activity could spill over to the adjacent region of the chromosome. Some evidence is available that the retrovirus can be sent to specific, harmless regions of the chromosomes. Nonetheless, it is a risk worth considering.

For all the hurdles yet to be surmounted, gene therapy has

already begun. It is sometimes said that medicine is lagging too far behind biomedical science. While scientists explain in fine details the very processes of life, there seems to be no tangible evidence of this progress in the clinics and hospitals, no supply of new miracle drugs to rid ourselves of the formidable diseases we face. But good applied science in medicine requires a broad and deep understanding of the basic facts about disease. Not until this is achieved can we hope to design a direct and decisive remedy. The record of medicine in this century bears this point out. The application of gene therapy is the most recent example of the marriage between the knowledge of the underlying disease mechanism and the technology of basic science.

The correction of ADA deficiency in children afflicted with SCID signalled a new era in medicine. SCID is a rare disorder and while it appears an obvious choice for this new mode of therapy, because the genetic defect is straightforward, the potential of gene therapy is much broader. In today's roster of important illnesses are the cardiovascular diseases and cancer. The possibility of preventing these or turning them around by the manipulation of genes before they run their destructive course is not just a theory being discussed at scientific conferences, but a therapy that has actually begun in medical clinics.

It is a totally new world.

BIBLIOGRAPHY

Chapter 1 Virchow's Cell

Burke, J., 1985. "What the doctor ordered." In *The day the universe changed*. British Broadcasting Corporation, London.

Handley, W.S., 1931. *The genesis of cancer.* Kegan Paul, Trench Trubner and Company, London.

Holmes, G.W., Daland, E., Warren, S., Simmons C., 1940. *Cancer: A manual for practitioners.* Boston, Massachusetts.

Malkin, H.M., 1990. Rudolf Virchow and the durability of cellular pathology. *Perspect Biol. Med.* 33:431-443.

Rather, L.J., 1978. *The genesis of cancer: A study in the history of ideas.* John Hopkins University Press, Baltimore.

Shimkin, M.B., 1977. *Contrary to nature.* U.S. Department of Health, Education and Welfare, Washington, D.C.

Triolo, V.A., 1965. Nineteenth century foundations of cancer research. Advances in tumor pathology, nomenclature, and theories of oncogenesis. *Cancer Res.* 25:75-106.

Chapter 2 The Language of Life

Augenlicht, L.H., 1992. Gene structure, function and abnormalities. *Cancer* 70:1671-1684.

Avery, O.T., MacLeod, C.M., McCarty, M., 1944. Studies on the chemical nature of the substance inducing transformation of pneumococcal types. Induction of transformation by a deoxyribonucleic acid fraction isolated from Pneumococcus Type III. *J. Exp. Med.* 79:137-158.

Beadle, G., 1946. Genes and the chemistry of the organism. *Am. Sci.* 34:31-53.

Crick, F.H.C., Barnett, L., Brenner, S., Watts-Tobin, R.L., 1961. General nature of the genetic code for proteins. *Nature* 192:1227-1232.

Crick, F., 1970. Central dogma of molecular biology. *Nature* 227:561:563.

Darnell, J.E., 1985. RNA. *Sci. Amer.* 253:26-36.

Dunn, L.C., 1965. *A short history of genetics.* McGraw-Hill Book Company, New York.

Felsenfeld, G., 1985. DNA. *Sci. Amer.* 253:58-67.

Nirenberg, M., 1973. "The genetic code." In *Nobel Lectures - Physiology or Medicine 1963-1970.* American Elsevier.

Smith, E.L., Hill, R.L., Lehman, I.R., Lefkowitz, R.J., Handler, P., White, A., 1983. " The gene and its replication." In *Principles of Biochemistry.* McGraw-Hill Book Company, New York.

Watson, J.D., Crick, F.H.C., 1953a. Molecular structure of nucleic acids. *Nature* 171:737-738.

Watson, J.D., Crick, F.H.C., 1953b. Genetic implications of the structure of deoxyribonucleic acid. *Nature* 171:964-967.

Watson, J.D., 1968. *The Double Helix.* Atheneum, New York.

Chapter 3 Lessons at the Threshold of Life

Anderson, N.G., 1970. Evolutionary significance of virus infection. *Nature* 227:1346-1347.

Bishop, J.M., 1987. The molecular genetics of cancer. *Science* 235:305-311.

Butler, P.J.G., Klug, A., 1978. The assembly of a virus. *Sci. Amer.* 239:62-69.

Duesberg, P.H., Vogt, P.K., 1970. Differences between ribonucleic acids of transforming and non-transforming avian tumor viruses. *Proc. Natl. Acad. Sci.* 67:1673-1680.

Fraenkel-Conrat, H., Williams, R.C., 1955. Reconstitution of active tobacco mosaic virus from its inactive protein and nucleic acid components. *Proc. Natl. Acad. Sci. USA* 41:690-698.

Hardy, W.D., 1983. Naturally occurring retroviruses. *Cancer*

Investigation 1:67-83.

Hershey, A.D., Chase, M., 1952. Independent functions of viral protein and nucleic acid in growth of bacteriophage. *J. Gen. Physiol.* 36:39-56.

Rous, P., 1911. A sarcoma of the fowl transmissible by an agent separable from the tumor cells. *J. Exp. Med.* 13:397-411.

Rous, P., 1967. The challenge to man of the neoplastic cell. *Science* 157:24-28.

Stanley, W.M., 1935. Isolation of a crystalline protein possessing the properties of tobacco-mosaic virus. *Science* 81:644-645.

Temin, H.M., 1974. On the origins of RNA tumor viruses. *Ann. Rev. Genetics* 8:155-177.

Temin, H.M., 1976. The DNA provirus hypothesis: the establishment and implications of RNA-directed DNA synthesis. *Science* 192:1075-1080.

Todaro, G.J., 1975. Evolution and modes of transmission of RNA tumor viruses. *Am. J. Pathol.* 81:590-606.

Varmus, H.E., 1987. Reverse transcriptase. *Sci. Amer.* 257:56-64.

Chapter 4 Betrayal from Within

Barbacid, M., 1987. *Ras* genes. *Ann. Rev. Biochem.* 56:779-828.

Bishop, J.M., 1986. From proto-oncogene to oncogene. *Advances in Oncology* 2(4):3-8.

Croce, C.M., Klein, G., 1985. Chromosome translocations and human cancer. *Sci. Amer.* 252:54-60.

Doolittle, R., 1981. Similar amino acid sequences: Chance or common ancestry? *Science* 214:149-159.

Hollstein, M., Sidransky, D., Vogelstein, B., Harris, C.C., 1991. *p53* mutations in human cancers. *Science* 253:49-53.

Huebner, R.J., Todaro, G.J., 1969. Oncogenes of RNA tumor viruses as determinants of cancer. *Proc. Natl. Acad. of Sci. USA* 64:1087-1094.

Hunter, T., 1984. The proteins of oncogenes. *Sci. Amer.* 251:70-79.

Kingston, R.E., Baldwin, A.S., Sharp, P.A., 1985. Transcription control by oncogenes. *Cell* 41:3-5.

Klein, G., 1987. The approaching era of the tumor suppressor genes. *Science* 238:1539-1545.

Knudson, A.G., 1985. Hereditary cancer, oncogenes, and anti-oncogenes. *Cancer Res.* 45:1437-1443.

Land, H., Parada, L.F., Weinberg, R.A., 1983. Cellular oncogenes and multistep carcinogenesis. *Science* 222:771-778.

Lebowitz, P., 1983. Oncogenic genes and their potential role in human malignancy. *J. Clin. Oncol.* 1:657-662.

Linder, M.E., Gilman, A.G., 1992. G proteins. *Sci. Amer.* 267:56-65.

Malkin, D., Li, F.P., Strong, L.C., Fraumeni, J.F., Nelson, C.E., Kim, D.H., Kasse, I.J., Gryko, M.A., Bischoff, F.Z., Tainsky, M.A., Friend, S.H., 1990. Germ line *p53* mutations in a familial syndrome of breast cancer, sarcomas, and other neoplasms. *Science* 250:1233-1238.

Nowell, P.C., Hungerford, D.A., 1960. A minute chromosome in human chronic granulocytic leukemia. *Science* 132:1497.

Murphree, A.L., Benedict, W.F., 1984. Retinoblastoma: clues to human oncogenesis. *Science* 223:1028-1033.

Sager, R., 1986. Genetic suppression of tumor formation: A new frontier in cancer research. *Cancer Res.* 46:1573-1580.

Tabin, C.J., Bradley, S.M., Bargmann, C.I., Weinberg, R.A., 1982. Mechanism of activation of a human oncogene. *Nature* 300:143-149.

Varmus, H., Weinberg, R., 1993. *Genes and the biology of cancer.* Scientific American Library. New York, NY.

Vogelstein, B., 1990. Cancer - a deadly inheritance. *Nature* 348:681-682.

Vogelstein, B., Kinzler, K.W., 1992. *p53* function and dysfunction. *Cell* 70:523-526.

Weiss, R., 1983. Oncogenes and growth factors. *Nature.* 304:12.

Weinberg, R.A., 1983. A molecular bases of cancer. *Sci. Amer.* 249:126-142.

Weinberg, R.A., 1985. The action of oncogenes in the cytoplasm and nucleus. *Science* 230:770-776.

Weinberg, R.A., 1991. Tumor suppression genes. *Science*

254:1138-1146.

Wigler, M., 1986. The role of *ras* in oncogenesis. *Advances in Oncology* 2(1):4-7.

Yunis, J.J., 1983. The chromosomal basis of human neoplasia. *Science* 221:227-236.

Chapter 5 Survival of the Fittest

Baserga, R., 1981. The cell cycle. *N. Engl. J. Med.* 304:453-459.

Buick, R.N., Pollack, M.N., 1984. Perspectives on clonogenic tumor cells, stem cells, and oncogenes. *Cancer Res.* 44:4904-4918.

Fidler, I., Hart, I., 1982. Biological diversity in metastatic neoplasms: origins and implications. *Science* 217:998-1003.

Levine, A.S., 1984. Fruit flies, yeasts and *onc* genes: Developmental biology and cancer research come together. *Med. Pediatric Oncol.* 12:357-374.

Mackillop, W.J., Ciampi, A., Till, J.E., Buick, R.N., 1983. A stem cell model of human tumor growth: Implications for tumor cell clonogenic assays. *JNCI* 70:9-16.

Nowell, P.C., 1976. The clonal evolution of tumor cell populations. *Science* 194:23-28.

Nowell, P.C., 1986. Mechanisms of tumor progression. *Cancer Res.* 46:2203-2207.

Tannock,I., 1989. "Principles of cell proliferation: Cell kinetics." In *Cancer: Principles and practice of oncology.* De Vita,V.T., Hellman, S., Rosenberg, S.A. (eds.), Vol.1, 3rd edition. J.B. Lippincott Company, Philadelphia.

Chapter 6 Contrary to Nature

Bell, D.R., Gerlach, J.H., Kartner, N., Buick, R.N., Ling, V., 1985. Detection of P-glycoprotein in ovarian cancer: A molecular marker associated with multidrug resistance. *J. Clin. Oncol.* 3:311-315.

Cairns, J., 1975. Mutation selection and the natural history of cancer. *Nature* 255:197-200.

Fidler, I.J., 1984. Recent concepts of cancer metastases and their implications for therapy. *Cancer Treat. Rep.* 68:193-198.

Kartner, N., Riordan, J.R., Ling, V., 1983. Cell surface P-glyco-protein associated with multidrug resistance in mammalian cell lines. *Science* 221:1285-1287.

Liotta, L., 1992. Cancer cell invasion and metastasis. *Sci. Amer.* 266:54-63.

Schimke, R.T., 1983. Gene amplificatin and drug resistance. *Sci. Amer.* 243:60-69.

Schnipper, L.E., 1986. Clinical implications of tumor cell heterogeneity. *N. Engl. J. Med.* 314:1423-1431.

Chapter 7 The Fourth Frontier

Anderson, W.F., 1984. Prospects for human gene therapy. *Science* 226:401-409.

Anderson, W.F., 1992. Human gene therapy. *Science* 256:808-813.

Bains, W., 1987. *Genetic engineering for almost everybody.* Penguin Book, London.

Broder, S., 1991. Progress and challenges in the global effort against cancer. *J. Cancer Res. Clin. Oncol.* 117:290-294.

Culver, K., Cornetta, K., Morgan, R., Morecki, S., Aebersold, P., Kasid, A., Lotze, M., Rosenberg, S.A., Anderson, W.F., Blaese, M., 1991. Lymphotcytes as cellular vehicles for gene therapy in mouse and man. *Proc. Natl. Acad. Sci. USA* 88:3155-3159.

Culver, K.W., Osborne, W.R., Miller, A.D., Fleisher, T.A., Berger, M., Anderson, W.F., Blaese, R.M., 1991. Correction of ADA deficiency in human T lymphocytes using retroviral-mediated gene transfer. *Transplantation Proc.* 23:170-171.

Friedman, T., 1992. Gene therapy of cancer through restoration of tumor-suppressor functions? *Cancer* 70:1810-1817.

Gewirtz, A.M., 1993. Therapeutic applications of antisense

DNA in the treatment of human leukemia. *Proc. Amer. Assoc. Cancer Res.* 34:595.

Hock, R.A., Miller, A.D., Osborne, W.R., 1989. Expression of human adenosine deaminase from various strong promoters after gene transfer into human hematopoietic cell lines. *Blood* 74:876-881.

Mulligan, R.C., 1993. The basic science of gene therapy. *Science* 260: 926-931.

Ozturk, M., Ponchel, F., Puisieux, A., 1992. *p53* as a potential target in cancer therapy. *Bone Marrow Transplantation* 9:164-170.

Rosenberg, S.A., Aebersold, P., Cornetta, K., Kasid, A., Morgan, R., Moen, R., Karson, E., Lotze, M., Yang, J.C., Topalian, S.L., Merino, M.J., Culver, K., Miller, A.D., Blaese, R.M., Anderson, W.F., 1990. Gene transfer into humans - Immotherapy of patients with advanced melanoma, using tumor-infiltrating lymphocytes modified by retroviral gene transduction. *N. Engl. J. Med.* 323:570-578.

Roth, J.A., 1993. Use of retroviral vectors for the delivery of antisense constructs. *Proc. Amer. Assoc. Cancer Res.* 34: 594.

Thomas, L., 1974. "Medical lessons from history." In *The medusa and the snail.* Bantam Books, New York.

Weinberg, R.A., 1992. The integration of molecular genetics into cancer management. *Cancer* 70:1653-1658.